Building with Flint

A Practical Guide to the Use of Flint
in Design and Architecture

Building with Flint

A Practical Guide to the Use of Flint in Design and Architecture

David Smith

THE CROWOOD PRESS

Contents

Introduction		6
Chapter One	What is Flint?	7
Chapter Two	History and Usage	10
Chapter Three	Flint Architecture	27
Chapter Four	Flint Styles	48
Chapter Five	Construction Planning	80
Chapter Six	Tools to Use	120
Chapter Seven	Flint Knapping	128
Chapter Eight	Building a Wall	132
Chapter Nine	Support	147
Conclusion		156
Bibliography		157
Index		158

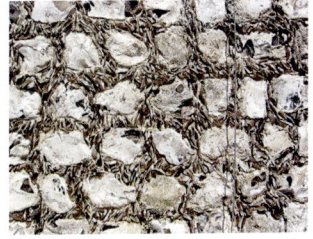

This book is intended to be used as a practical resource of interest both to those working in the professions of architecture, building and design and to anyone with a curiosity and enthusiasm to discover more about this material, which has come to represent archetypal qualities of durability, versatility and strength.

In the evolving relationship between humans and the natural world, the provenance of flint as a resource is, arguably, unparalleled. Its continuing use today is simply another link in a long chain of association that can swiftly transport the craftsman back to his ancient ancestors.

In historical terms, human beings were relatively quick to discover the usefulness and versatility of flint. It offered itself up readily, rising to the surface of the land. But perhaps part of the enduring fascination that flint holds for humans is that it does not easily give up its secrets.

Within the stone are held the solutions to many of the fundamental challenges of early human life – how to successfully hunt for and prepare food, protect oneself from enemies and provide heat and shelter for self and tribe. But flint provides no easy answers. It requires observation, investigation, technical learning, and dedication to practise its use. It could be fancifully imagined that its availability and the tantalising alchemical promise of it, has challenged humans to organise their efforts more methodically, systematically, and experimentally in the manner of mathematical calculation and scientific enquiry. It rewards creativity too, the ability to closely observe and imaginatively predict; the artist's skill in recognising the potential of a chance discovery.

Flint has been used for a number of diverse purposes, ranging from weapons, hand tools, farm equipment, jewellery and personal adornment, ceramics, glass making and even musical instruments. However, it is perhaps not surprising that the potential of flint is currently enjoying something of a renaissance within the practice of architecture, a field that aspires to combine and balance function with aesthetics.

Although touching on some of the above uses of flint, the majority and practical elements of this book predominantly focus on the uses of flint as a material for construction. Its contents and terminology are by no means exhaustive: it would be impossible to include every single example of flint use and style of architecture. Throughout the book there are examples of other applications that demonstrate some of the under-explored qualities of this versatile stone. Does it yet have solutions to offer to some of the contemporary environmental problems of sustainability that we face today?

Flint has historically been a material available to everyone, it is 'the crop that never fails'. It is hoped that this book will help to create a spark of excitement and interest in its readership and to encourage the continuation of the timeless tradition of learning from this unique stone, the story of which is so fundamentally bonded with our own.

Chapter One

What is Flint?

Formation

Flint (SiO_2 – silicon dioxide) is a microcrystalline sedimentary material composed of silica. Its very pure silica content is made of quartz, opal and chalcedony. It formed in the chalk deposits of southern England and western Europe during the late Cretaceous period, 65 to 95 million years ago. The silica in flint is derived mainly from the skeleton remains of marine organisms and micro-organisms such as sponges, diatoms and radiolarians. Some silicon spherules may have precipitated directly on the seafloor.

Silica from various sources is buried with the background chalk algal coccolith sediment. With depth of burial, bacterial systems working on the organic matter in the chalk cause changes in the chemistry within the seabed and these changes cause local acidic conditions where the chalk carbonate is dissolved and silica is precipitated, forming the beginnings of flint. The chemical change occurs at the 'redox boundary' (reduction/oxidation boundary), which is parallel to but below the seafloor and remains stable in cycles. Hence, flint bands form in layers parallel to the seafloor but not on it. The easiest places for the chemical reactions to take place are in the least compacted chalk, which is in the animal burrow-fills, and this is where the flint starts to form, retaining the shape of the original burrow systems.

The process in the formation of flint in chalk is called diagenesis. This refers to the chemical, physical and biological changes that occur in rock as it forms and metamorphoses. Certain periods of time brought changes including climate cycles, biological cycles and sea-level fluctuations. These changes in conditions created a warming of the sea. Unable to survive during these events, certain marine organisms died and their skeletal remains fell to the seafloor. The gases caused by this bacterial activity, together with the oxygen levels in the seabed, created strong acidic conditions resulting in their silica content to dissolve in the water contained in the chalk. The silica then builds up and consolidates in shallow voids, as well as a network of cavities previously formed in the seabed by worms, molluscs and crustaceans. The silica gradually crystallised and hardened to form flint.

The flint band frequency, density and size depends on the climatic conditions, the depth of the sea, the numbers of decaying organisms and their silica content. The horizontal bands of flint that can be seen today between the white chalk cliff faces of southern England show that there was more than one episode and climate cycle that created the optimum conditions to enable flint formation. In general, the younger white chalk strata contain larger formations of flint. Some of these exact same flint beds can be identified and traced as the same band from the south coast to a hundred miles away on the East Anglian and Yorkshire coastlines.

Types

Although it has its own geological classification, flint in the United Kingdom is often confused with the substance known as 'chert'. They are both composed of the same mineral properties but have different grain sizes and porosity. The main types of flint are termed field flint, chalk quarried flint, gravel pit flint and cobbles (sometimes known as sea flint).

A detail of a flint band in chalk cliff, East Sussex.

Appearance

Flint can be found in different shapes and sizes. The shape of its formation was dependent on the terrain and conduits on the seabed. As well as animal burrows, these cavities included fissures, fractures and voids in the chalk. The main shape types formed are named as nodular, tabular, tubular and sheet flint. An unusual exception to this is the paramoudra, sometimes known as potstones or sea pears. These are trace fossils that are created by microcrystals of silica forming concentrically around a fragment of fossil and are often found in coastal locations, particularly around West Runcton, Norfolk. They can vary in size and be found up to 2 metres in diameter.

The outside of a flint is called the cortex, rind or skin. This is the transition between the flint forming and the incompletion of silicification. This can come in varying thickness, textures and in several colours. All of these can be influenced by the nature of the original formation, texture and the staining of mineral content of the parent rock, subsequent geological movement and exposure to the elements. Cortex colours can vary from white, brown, blue, green and grey.

The inside of a flint is called the core. Although flint has a high silica purity content, the small remaining impurities and absorption of mineral salts can affect its qualities and appearance in colour and patination. These trace elements can include iron, calcium, potassium, magnesium and sodium. Time exposed to the elements can be a huge influence on the colour and patination. In some instances, if split, the core will react with the change in chemical elements and can have an immediate change of colour. Due to an optical effect of both the high density of the silica reducing light movement and the cortex preventing light access, it is impossible to view the true colour of the flint unless either a thin flake or without any cortex remaining. Flint core colours and patination can vary from white, brown, blue, green and grey.

Occasionally, it is possible to find fossils in the flint itself. This is either due to skeletal remains in the cavities before silica build up, or animals entering the cavity whilst the silica is still in fluid form and before silicification has taken place. Formation of flint is post-depositional. As flint forms it replaces anything that is present in that band.

Paramoudra on a quarry floor. Also known as potstones or sea pears.

Geography

Flint can be found throughout the United Kingdom but is more prolific in the southern counties of England, from Devon to Yorkshire. Subject to erosion of chalk and glacial movement, flint can be found in numerous locations including on the surface of fields, in chalk cliffs, on the beach, in riverbeds and in gravel pits.

'Flint' in Many Tongues

Afrikaans:	Vuurklip	Hungarian:	Kova
Albanian:	Sterrall	Irish:	Cloch thine
Armenian:	Khijabagi	Italian:	Silice
Bulgarian:	Krem'k	Norwegian:	Flint
Chinese:	Huo shih	Polish:	Krzemien
Danish:	Flint	Romanian:	Cremene
Dutch:	Vuurstteem	Russian:	Kremen
Eskimo:	Kukiksaut	Spanish:	Pedernal
French:	Silex	Swedish:	Flinta
German:	Feuerstein	Turkish:	Calmak tasi
Greek:	Pyrites	Welsh:	Callestr

(**Source:** Shepherd, W., *Flint, Its Origins, Properties and Uses*)

The Tale of Flint Jack

In the Victorian era flint archetypes became sought-after by amateur collectors, providing opportunities for forgers to exploit. The name most associated with these practices is that of 'Flint Jack', who appears to have started out in North Yorkshire as Edward Simpson, the servant of a minister and geologist. It might be assumed that this is where he began to gather an interest and knowledge of fossils and archaeological artefacts. With his own necessarily unreliable narration at its core, it is difficult to separate fact from myth. Accounts of the times credit him with a prodigious output of fake antiquities said to be acquired from Flint Jack (also known by a number of other names including 'Bag o' Bones', The Old Antiquarian, Stone Jack and Fossil Willy). His skill in knapping flint objects fine enough to capture the imagination of credulous local historians and enthusiasts cannot be doubted. However, his own creative imagination was his undoing on occasion, and the authenticity of some particularly finely wrought fishhooks and a flint comb notably raised suspicions when assessed by museum experts.

With an increased awareness of the work of forgers, collectors became less easily duped and the fortunes of those who had mastered this duplicitous art declined. Some found more respectable occupations. 'Flint Jack' continued to live as an itinerant trader scratching a living, in ill health and descending into alcoholic addiction.

The last official record shows that a magistrate's court found him guilty of felony in 1874 and he was sentenced to one year's imprisonment, after which only apocryphal stories remain. His legendary reputation was assured following the publication of an article in a Yorkshire newspaper in 1888, which claimed to chronicle his unscrupulous activities and condemned him as an irredeemable rogue.

Flint Jack.

The working of gunflints was a large and lucrative industry. As a result of this, more people were able to knap and the availability of flint for use in architecture increased. Flint, or a particular quality of flint, again became a very important commodity. The difference was that now it was worked by iron and steel rather than antler and stone.

Origins

There is no clear evidence showing when and where the gunflint industry started. On a small scale it may have been as early as the start of the seventeenth century. However, it quickly became a lucrative industry in France, England, Belgium and Germany. It is clear that the industry evolved as the arms industry and gun design advanced. Individuals were looking for more reliable ways of firing weapons, compared to the matchlock method (using a burning rope). During this time, the latest and most popular weapons (due to speed of fire and reliability) needed flintlocks. These included the Spanish 'miquelet' or the earlier German 'schnapphahn' (appropriately known as 'the pecking hen'), the Dutch 'snaphance' and later the English 'flintlock'.

The gunflint industry reached its peak during the Napoleonic period. At this time, there are records showing up to 800 individuals working as part of the flint industry in France. The Department of Loire et-Cher contained rich deposits of flint, as detailed in a museum at Meusnes. Alongside gunflint production, many iconic flint buildings were built by Napoleonic prisoners-of-war during this period. This may be due to the access of labour, and high-quality workable flint, plus the architectural flint fashions of the time. However, the possible flint knapping skills of the French labourers are also likely to have played a role.

Meanwhile in England, new quarries were opening up across many southern counties to match the extensive demand and search for better-quality flints. These included Wiltshire, Essex, Kent and Norfolk. Numerous new gunflint practices opened up near these resources to cope with the increasing demand. There are also records that show that the skilled labour who were needed to make gunflints were peripatetic workers who would travel where their skills were needed.

Brandon knappers

It was, however, the town of Brandon in Suffolk that became the best-known home of the gunflint industry. There now remains little physical evidence in the town of the importance and magnitude that the

An English flintlock blunderbuss pistol with gunflints.

industry once enjoyed in everyday life. There are no longer the numerous knapping booths, flint storage yards and tool shops. In fact, only one remains – the rest have either been demolished or converted for other uses. The only clues remaining are a pub sign, a few unusual decorative panels and cottages made of gunflint quartering waste and round gunflint cores. There is some debate on why it was the village of Brandon that was such a centre of flint production. It is no coincidence that Brandon is close to Grimes Graves, one of the largest Neolithic flint mines in the country. Both Neolithic man and the flint knappers of the gunflint industry recognised the outstanding quality of the local flint, particularly the nature of the floor-stone flint. However, with other areas in other parts of the country offering similarly high-quality flint, there still begs the question why Brandon? Yes, there was the perfect match of both skill base and a source of good quality flint there that perhaps other places like Fleet in Essex had one, but not the other. But there are implications of family connections between local landowners and the Board of Ordnance. So, perhaps therefore it is no coincidence that in 1790 Philip Hayward of Bury St Edmunds, who was awarded a government contract of 100,000 gunflints, subsequently secured Brandon as a gunflint centre of such renown.

After this, Brandon flint masters had the sole contract to supply the British army with gunflints from 1804 onwards. The black Brandon gunflint was valued more than any other source due to its quality and reliability – so much so that less scrupulous traders would engage in the dubious practice of 'spotting' or 'lamp blacking' gunflint: darkening the gunflint and hiding any imperfections to guarantee a sale or to obtain a higher price. At the time the military would not purchase any flint with chert inclusions within a quarter of an inch of the edge of the flint. It was also said the Brandon flint was 'jackdaw-coloured, free of impurities and of good running quality'. The best Brandon gunflint would last up to 50 shots. This was allegedly much more than the inferior gunflints sourced and supplied in other parts of the country. By 1804, the Board of Ordnance commissioned Brandon flint makers to supply 360,000 gunflints a month. By 1813, there were 160 recorded knappers and diggers. Although it was a male-dominated trade, records reference a female gunflint master named Elizabeth Grief. There are also records showing other female manufacturers and listed subscribers, but it is hard to know what their actual roles were. Family names that have become synonymous with the gunflint industry began to emerge: the Snares, the Carters, the Fields and the Edwards families. They were all family businesses, with specific reputations. Allegedly, it is said that skilled gunflint makers could identify their own craftsmanship by touch. Makers were not allowed to sub-contract any of the required gunflint orders. Although there are records showing that the skill levels and quality of gunflints varied, each maker prided themselves on the quality of their work. Orders were made to a specific gunflint maker depending on which of the 23 types of gunflint was required. As a result of this skilled craftsmanship, Brandon gunflint was shipped around the world for over 200 years, from 1790 to 1996.

As at Grimes Graves, flint was mined rather than quarried in Brandon. There are records showing up to ten main flint mine areas that have supported the Brandon gunflint industry over the years. These include Icklingham, Weeting and Santon – but probably the most famous and most productive of those areas was Lingheath. Here numerous shafts were dug just big enough to work the flint loose and transport it upwards. The layers of the flint strata would have been dug and brought to the surface. No machinery or mechanical means were used to extract either flint or chalk: everything was done by hand. The layers of the flint strata consisted of topstone, wallstone and floorstone. Shaft depths could be up to 40 feet (12.2m) deep depending on the level of the bottom floorstone. Mining was not exclusively for supplying the gunflint industry. No flint went to waste, with the less workable topstone and wallstone produced when digging a shaft going to the local building industry. However, it was higher-quality flint floorstone that the diggers were in search of exploiting. This flint was much more predictable and workable. Furthermore, it would have received more money per weight. It is for this reason that most galleries were only dug at the floor-stone level.

Mining

To reach any flint, a vertical shaft would have to be dug through the topsoil and down into the chalk. It would take around one week to dig a single shaft. Shafts were dug surprisingly close to one another, but skilfully enough that they did not encroach on each other or have dangerous structural issues. They would always be dug at an angle to aid the lifting process and to reduce any risk of falling debris. The average shaft could produce up to six or seven months of flint mining. At various stages down the shaft would be toe holes and platform steps. These would be used as staging posts for tools and to bring the heavy flint up to the surface. The ledges would be dug on three sides with the fourth side used to extend the main shaft. This system of digging at an angle or slant with a series of ledges was (and still is) known in East Anglia as 'bubberhutching on the sosh'. The term 'soshed' is comically still used locally to refer to a tipsy character who may be walking at an angle or 'on the sosh'.

Any mining would be carefully performed to reduce any excess chalk waste and save unnecessary time and labour. It was hard physical work. Most diggers worked either alone or sometimes with one other, who would help to bring either spoil or flint up to the surface. At Lingheath flint mine, records show that digging for flint was male dominated, although as with the gunflint making there are written records showing a couple of female diggers. With the exception of the shaft size and shape (smaller and less vertical), and small changes in technology, the process of mining for flint did not change much from the time of the Neolithic men at Grimes Graves. In alignment with the flint strata, horizontal galleries would also be dug. These would enable the digger to work and remove the flint that had been geologically formed at that level. For safety reasons, diggers would usually begin working the bottom floor stone galleries and gradually move up to the higher top stone galleries. This was a slow process, as it involved hand tools and very cramped working conditions.

Tools to remove the chalk and extract the flint were basic: a pronged pick, a spade, a crowbar and a hammer. The crowbar was used to dislodge the flint and the heavy hammer was used to break up any large nodules that were either too heavy or large to manoeuvre through the galleries or lift up to the surface. In contrast to the miners of Grimes Graves, the Lingheath diggers would excavate the galleries above the flint strata to remove the flint nodules. This may have made it harder to dislodge the flint, but it did reduce the excess chalk spoil. It is often assumed that for the Neolithic miners of Grimes Graves, excess spoil was less of an issue. Not only were the shafts to remove the spoil much bigger, but it would appear that the diggers worked as a group.

Despite the reflective qualities of chalk, the limited light in galleries and shafts meant that digging would have been undertaken by candlelight. Candles would also have been used as timepieces, as the repetitive tasks and absence of daylight made it hard to keep track of time. Most miners would use two candles a day. When a miner entered the shaft at the start of the day they would light their first candle. When that candle had burnt down to half its size, the miner would return to the surface for a sip of tea and some clean air before returning to the shaft. A second trip back to the surface would occur when the first candle had completely burnt down. This would be time for the digger's 'docky' (midday meal). When the second candle had burnt down completely, it would indicate the end of the working day.

Mined flints were measured by the 'jag' or cartload. A jag was usually measured by eye and not weight (however, it was equivalent to around a ton in weight). If they reached a good stratum, a digger could produce between three to four jags a week. A jag would

Pony Ashley mining flint at Lingheath, Suffolk.

be lifted up to the surface and placed in a crescent shape around the shaft. Some records indicate that flint stored at surface level would be protected from the elements with branches of Scots fir. Flints of different strata quality would be kept in separate piles due to their differences in value. Floor stone was sometimes almost worth twice the value of the top stone. This was because a jag of better-quality flint could produce almost double the amount of gunflints (around 12,000) than a poorer quality flint (around 6,000).

Quartering, flaking and knapping

Gunflint makers were in charge of sourcing and preparing the flint. It was their job to transform the raw material into either gunflints or knapped flints for the local building industry. They were stationed in numerous booths, sheds and yards in Brandon itself. The flint would have been transported by cart from the shaft to the knapper's yard. For practical and health reasons quartering was often completed outside, whereas the flaking and knapping was done in booths and sheds. The sheds were often cramped and small with limited natural light. The numbers of people working in a single workshop changed over time from three or four to one or two as the industry declined. The working conditions were poor, as

any ventilation would be blocked to reduce draughts blowing into the booth. This created a dangerously dense atmosphere of fine flint dust particles. In the early days of the industry little or no precautions would be taken. Later on in time and more aware of the risks, to reduce dust inhalation, knappers would either tie wet sponges and cloth around their face or drink beer. Unsurprisingly, both interventions only briefly prolonged the almost inevitable development of silicosis. The average life expectancy for knappers was only 44 years.

In the hands of the knapper, the flint went through a three-stage process: quartering, flaking and knapping. This was an editing and sorting process. A skilled knapper could accurately predict the quality of an unworked flint by feel or sound (for example, the ring it makes when hit with a hammer). However, often it is not until the flint has been opened up and inspected that the potential of the nodule can be understood. To reduce the risk of wasted labour, some poorer-quality nodules would have been put aside for the lesser demands of either the building industry or used for road building. These flints would have been graded into 'unworked', 'rough face' and 'snapped'. In addition, flints of a good workable quality would have been occasionally set aside for fine-gauged square architectural work by a gun-flint maker.

Quartering involved breaking down a large nodule into a more manageable size; this could be done by a knapper of any skill level with a large lump hammer. Depending on the flint size and knapper preference, the hammers varied in shape (square or hexagonal) and weight (3 to 7 pounds). The flint nodule would be placed on the knapper's knee. Following this, the top corner would be struck at a 45-degree angle. As well as making the nodule more manageable, knapping would open the flint up, exposing the flint 'face' and producing a platform for the flaking process to begin.

The flaking process relied on skilful knappers called 'flakers'. Similar to quartering, different hammers were selected depending on the flake size and gunflint required. Each individual firearm required a custom-sized gunflint. Normally a hammer with a square end was used to hit the outer part of the flint and create flakes of material. It must also be noted that French knappers were the first to use this unique 'platform' technique. The use of a square-head hammer meant longer and predictable flakes were produced. These resulted in less trimming and often more gunflints per flake. Though it is not known how English knappers acquired the French technique, it may have been learned and shared by travelling flint knappers. This plays into the flint knappers' myth that if you were ever captured in a war, proving your knapping abilities could save your life. Records show that the 'platform' technique was predominantly used in Norfolk and Suffolk, whereas the 'wedge' technique remained in Wiltshire and Essex.

The flaker would gradually work around the circumference of the nodule until the flint became unworkable. The unworkable matter was known as the core and would be discarded or set aside for the building industry. Depending on the quality of the flint, a skilled flaker could produce up to 6–7,000 flakes a day. Flakers created numerous terms for working the flint. These include a 'wrung' describing a flake with a twist in it, and a 'bruckly', a flake that is not easily worked.

The final part of the process was working the knapped flakes into gunflints. To do this, the flaker would place the piece of flint on a small anvil; then,

Flint quartering. William 'Billy' Carter, at the Carter's workshop on London Road, Brandon, c.1900.

by using a long, flat hammer (similar to a file in shape and weight), and placing the flint on an anchored vertical metal stake, the flake would be trimmed down to the finished product. Depending on the desired size of gunflint, and size of flake, sometimes four or five could be produced from a 6-inch (15cm) flake. Using good quality flakes, a skilful knapper could produce an average quota of around 300 gunflints an hour or up to 2,000 gunflints a day.

Saturday would be the knapper's counting day, or 'telling day', as it was locally known. A 'teller' would be someone responsible for counting gunflints: this was completed by pulling to the side three finished gunflints with the right hand and two with the left hand. A group of five was known as a 'cast'. Knappers could count up to 20,000 gunflints an hour using this method. Each cast was then placed into a hessian bag. Gunflints were generally bagged 'per mile' or a thousand before being placed into a barrel, ready for shipping. Knapping never took place on a Monday. As illiteracy was rife within the gunflint industry, knappers had their own numbering system for counting, recording and invoicing. These symbols would be made up of strokes, dashes, uprights, diagonals and crosses.

The demise of the gunflint industry

In 1815, following Napoleon's defeat at the Battle of Waterloo, gunflint production slowed. This was mainly due to ample stockpiles, reduced conflict, and the development of percussion weapons. At the time, the English Board of Ordnance did not renew some of the previous gunflint contracts (by 1839 the military only used percussion weapons). The business for the gunflint makers did not stop, but it was the start of the decline. This period also marked the start of turbulent times for Brandon and the surrounding areas. Poor working conditions and rising unemployment in the gunflint and agriculture industries, combined with soldiers returning home to no work, created a period of discontent. In 1816, anger and violence finally erupted into the Agrarian or Swing Riots of East Anglia.

Later in 1837, a group of Brandon knappers formed the Brandon Gunflint Company to control supply and prices, and attempt to consolidate a dwindling gunflint industry. This rebranding allowed the knappers to pool their skills and knowledge. However, a lack of business and poor management meant the company folded in 1849. Ironically, just four years later the Crimean War temporarily increased the demand for gunflint again. There are records of the Turkish Government ordering 11,000,000 carbine flints. Though new gunflint markets were emerging in Africa, India and the West Indies, by 1868 there were only 36 gunflint knappers and diggers left. Ten years later this number dropped to 26.

When the gunflint industry waned, knappers turned to lime burning, farming, architecture and even 'strike a lights'. By 1907, records show that there were only seventeen knappers and five diggers still working in Brandon. A combination of lack of demand and World War I with the exodus of many potential apprentices finally put a stop to Brandon's gunflint industry and the regular passing on of skills to the next generation. In 1950, this was reduced to just six flint knappers. These remaining knappers were producing around 40,000 gunflints a week for the African market and architectural flint for the booming post-war building industry.

In the 1960s, Fred Avery was the last person to be trained in Brandon as a flint knapper. By the 1970s and 1980s, the only flint worked in Brandon was gunflint for re-enactment societies and architectural flint for the building industry. Fred Avery was producing 75,000 flints in 1980 and 50,000 in 1995, providing him with part-time employment. Since his untimely death in 1996, there are no knappers remaining in the town itself – just a small museum exhibiting their paraphernalia and a couple of highly-skilled flint knappers based within 10 miles (16km) of Brandon who supply the building industry, archaeological artefacts and still the occasional gunflint.

With regard to quarrying, the last flint to be removed from Brandon's Lingheath mine was in the 1930s. Records show that 'Pony' Ashley was the last

Bill Basham's Flint Necklace

In the 1920s and early 1930s, Robert W. 'Bill' Basham was a well-known Brandon gunflint maker of skill and ability. In his spare time he made both a necklace and an alphabet out of flint. The alphabet took him two years to make. Sold for £10, it can now be seen at the Ancient House Museum, Thetford. In common with many of his peers, he died of silicosis at the young age of 38.

Robert W. 'Bill' Basham's flint alphabet. It took him two years to make and was sold for £10. It is on display at the Ancient House Museum, Thetford.

remaining miner making a living from digging up flint from Lingheath. He was remembered for always being covered in chalk. Similar to arms production, mining had also seen technological advances, meaning large-scale mechanical open pit quarrying was more profitable and economical than small-scale mining. Nowadays, there is little visual evidence of the previous century of digging and mining. The East Anglian countryside that was once riddled with shafts, jags and spoil-heaps is now a typical landscape of ploughed fields. Only on a long summer day when the sun is low, you might see the vague outline of a redundant shaft. Settlement in and around the back-filled mines has created odd circular impressions in the earth. These faint lines fail to convey the diligence and resilience of flint diggers at Lingheath: ghostly figures working the landscape covered in chalk and dust.

The Ceramic Industry

Since the early eighteenth century, flint had been extensively used in the ceramic industry. However, changes in manufacturing techniques and the development of other modern alternatives mean this is no longer the case. For many, it is a surprising fact that flint (in powder form) has been used to strengthen the body of clay and used in glazes. You may be currently drinking from a cup or eating off a plate that contains flint.

It is impossible to pinpoint exactly when flint was first introduced to the pottery industry. There are written records that John Dwight of Fulham patented the use of flint in pottery between 1671 and 1684. Small-scale flint mills were also in use at the time to supply the glass industry, although the first record of large-scale flint usage in the English ceramic industry was made in the early eighteenth century.

In the mid-seventeenth century, tea and coffee arrived in Britain with travellers returning from the Far East. By the early to mid-eighteenth century, these beverages had increased in popularity and were readily available. They were no longer being served in exclusive tea and coffee houses, but now consumed in people's homes. This transition coincided with the emergence of the middle classes, a wealthier socio-economic group of people with more leisure time, aspirations and money to spend. As a result, there was demand for better ware that matched the porcelain and white china that was being imported with the raw ingredients. This preferred ceramic was thinner, whiter and more translucent – a strong contrast to the yellow and red earthenware that was being produced in the country.

This earthenware was first introduced in the area of Stoke-on-Trent, aptly named 'The Potteries'. With a local abundance of clay and good quality coal, these six hamlets (that gradually amalgamated into one) already had a thriving pottery industry. Due to the increasing popularity of pottery, it had already evolved from the small-scale farmstead production of vessels to a larger-scale manufacturing process.

The introduction of flint was just what was needed to match the demand of higher-quality imported porcelain. Flint was heated up, crushed into powder form and added to the clay and glazes. This strengthened

The Story of John Astbury Introducing Flint to the Potteries

John Astbury was travelling by horse to London. On his journey he reached as far as Dunstable, where a local hotelier noticed he was having issues with the health of his steed. To rectify the horse's eye issues, the hotelier duly picked up a piece of flint from the ground, placed it into a fire to make it into powdered form and blew it into the eyes of the horse. This discharged the excess moisture in the horse's eye and enabled Astbury to venture on.

It is hard to know if this story is true, and if so, how the flint would have actually helped the animal. Furthermore, it seems feasible that Astbury could have visited Fulham or became aware of Dwight's methods or that of the glass manufacturing process from frequent trips to London. It may well be a myth or an elaborate pretence, but Astbury is acknowledged for transforming the manufacturing process and addressing an issue that the pottery industry was faced with.

Flint added to clay and glazes transformed the 18th-Century English ceramic industry.

the clay body, thus reducing shrinkage, and produced a whiter, glassier finish. The finished product had an appealing glaze finish. With the addition of the imported china clay (kaolin) from Cornwall and parts of France, the name 'Creamware' was born. Coined by Astbury's son Thomas, the name was chosen to match the lead-glazed earthenware he was producing.

For its abundance and ease of sourcing, cobbles from the beach were the main source of flint. There are no records to show if either other types of flint were experimented with or used. It is clear that cobbles from the south coast of England and northern France would have been relatively easy to transport to the region. In 1766 this would have improved even more with the digging of the Trent and Mersey Canal. This provided a more direct and navigational route from the south coast to the Potteries, saving money and time. In 1836 Trent and Mersey Canal records show that in one year 30,000 tons of cobbles were shipped from Newhaven and Gravesend to the Potteries.

Flint mills and grinding

The increase in demand for flint as a component of improved wares led to the introduction of flint mills. Initially, the mills were designed to transform flint from their raw form into a usable powder form for the demands of the booming ceramic industry. Originally, this task was completed by converting locally existing wind- or water-powered grain mills; high demand meant new purpose-built mills were created. Many of the larger potteries would have their own mill.

Before milling, the process first involved roasting the flint in its raw state. This was completed by loading purpose-built kilns with alternating layers of flint and coal. Approximately 1 ton of fuel was needed per 20 tons of flint. The kilns would then burn for approximately eight to sixteen hours to calcify the flint. This would then be washed to remove any fuel residues and subsequently crushed, pounded and sieved to remove any unwanted over-sized particles. It was a dry process and produced clouds of flint dust particles that were highly deleterious to health.

In 1726 Thomas Benson developed the wet pan grinding method. He had recognised the unacceptable working conditions that dry pounding caused. It was creating health issues for the workers similar to those of the flintlock industry. Although the health complications caused in the flintlock industry were called 'knapper's rot', in the pottery industry it was known as 'potter's rot': a different vernacular term for silicosis of the lung.

Benson's earlier attempts of wet grinding via an iron ball process were less successful. The residue of iron deposits made its way into the clay body, causing issues with the firing and finish. In 1732, Benson's process evolved through the introduction of heavy siliceous stones that ground the calcined flint. In the early days of wet grinding, the granite millstones of the grain industry were utilised. In search of improved methods and materials to make the best pottery, even the source and selection of grinding stone evolved. It is recorded that Josiah Wedgwood would only use Derbyshire chert for his grinding stones. This was due to old granite millstones leaving unwanted residue of black granite particles in the pure white ground flint powder. In the late nineteenth century for efficiency and particle size, pan mills began to be replaced by wet ball mill methods, although pan milling continued into the 1970s.

Once ground, the fine particle flint would be held in suspension and be placed in settling tanks. Here, as the flint settled the water would gradually run off until a thick suspension or 'slop' remained. The flint could be shipped to the customer in a 'slop' form or moved to the drying beds to produce a cake, which reduced transport costs.

Coal-fired flint kilns continued to be used up to the mid-twentieth century, when they started to be replaced by gas-, oil- or electric-fired kilns. There are no longer any active flint kilns in Stoke. However, on a much smaller level, flint still continues to this day to be processed in the Potteries area for the ceramic industry. One factory still imports calcified crushed flint from northern France and wet-grinds it to supply the ceramic industry. Now transported by road rather than canal, once processed it is then shipped in liquid form. It is, however, unknown how long this will last,

The wetstone grinding process. Heavy siliceous stones ground the calcined flint to powder. Etruria flint and bone mill, Stoke-on-Trent.

Furlong mills. One of the last remaining flint mills in Stoke-on-Trent that still supplies the ceramic industry.

as the need for flint in the pottery body and glazing process is now often replaced with alternatives such as quartz sand and alumina.

Cobbles in raw or calcified form from the United Kingdom are no longer used in the ceramic industry. This practice ended in the 1980s, with the last coming from the Northfleet area. Most now is imported in crushed calcified form from various sources on the north French coast. This comes in up to half a dozen grades.

English flint is still sourced on a smaller scale. At one point it was regularly exported to America for various industries, including the manufacture

Flint glass beakers.

Early Flint Glass Recipes

'A form of *cristallo* using powdered flints was being made as early as the fourteenth century before the intrusion of the Black Death. Potash from grape vines as well as soda was used in formulating the batch where appropriate. So were extraordinary materials such as sheeps' shin bones'.

(**Source:** *Ricette vetrarie del Rinascimento* by C. Moretti and T. Toninato, which transcribes a previously unknown mid-sixteenth century manuscript of Venetian glass-making practices and recipes.)

of toothpaste. Now small qualities are crushed and used for decorative floor finishes, and some are sold in raw form for water filtration here in the UK, or exported to the Asian market for grinding processes in manufacturing.

Flint Glass

One of the main components in the process of making glass is silica. In the seventeenth century due to its availability and high purity qualities, flint was used as the main source of silica. In fact there is evidence from a sixteenth-century manuscript of Venetian glass-making recipes that flints had possibly been used in glass making as early as the mid-fourteenth century.

During the seventeenth century, flint silica was much sought after, due to the highly refractive qualities it gave the glass. It was commonly used in the production of scientific equipment such as prisms and optical lenses for telescopes and microscopes. The abundance of this raw material that could be easily shipped up to London from the south coast of England, attracted glass-makers from the continent. The quality and quantity of flint available helped England become one of the leading glass manufacturers of the time. However, whatever the quality of flint, the glass-making process did have its problems. Flint glass was still too unstable, liable to break and would often fade or become opaque. Frustrated by the unpredictability of this process and the wastage involved, in 1673 London glass-maker George Ravenscroft decided to experiment with the glass-making recipes and processes. He discovered that with the additional use of lead oxide instead of the previously-used potash he could produce a glass that was not only more stable, but also with increased refractive qualities. This was called 'flint glass' – now more commonly known as 'lead crystal' or 'crystal glass'. To mark and identify quality and in reference to Ravenscroft, the early examples would have a raven's head on them.

By the late seventeenth century further improvements to the glass manufacturing process had changed the way flints were used. The flint mill was invented. A possible precursor to the flint mills used en masse later in the ceramics industry, flints were now burnt, crushed and milled. It is hard to know how long this process lasted, however records do show that by the early eighteenth century sand instead of crushed flint was being used more regularly. Whether it was to avoid the time-consuming milling process or whether this process could not keep up with demand, it is hard to know. By the mid-nineteenth century there is no evidence of flint being used by the glass-making industry.

Strike-A-Light and Steel Mills

Historically, lighting fires or lighting of some sort was for many a daily occurrence. Matches were not introduced until 1832. Previously there were numerous ways of creating a spark and lighting a fire. Striking flint with pyrite was one such popular method. Pyrite can often be found alongside flint in a sedimentary environment. Although reliable, in terms of spark control this was a pretty inefficient and painful method. In the sixteenth century the more efficient 'strike-a-lights' were invented and in common use. This was a small metal tool that when struck repeatedly with flint would produce a controlled spark. Directed towards tinder or other flammable material, this allowed a fire or candle to be easily lit. Up to 1970 it was a requirement within the Roman Catholic Church that at the start of the Easter Vigil and on special occasions the 'paschal candle' must be lit with 'new fire'. This would be stone, and traditional flint.

In the mid-seventeenth century, the steel mill was invented. This was a framed mechanism with a handle and thin steel wheel. When spun at speed and with a flint held against the wheel, it would produce a volume of sparks. Unlike the strike-a-light, the steel mill was invented to provide light and not light fires. In fact it was hoped it was the last thing it would do. Invented by Carlisle Spedding, its primary purpose was to replace the use of candles in more volatile areas of coal mines, and reduce the risk of igniting methane gas. The lightweight mill would be attached to the forearm and used when required. Ironically in 1755 Spedding died from gas poisoning in one of his own pits. By

The Flint Piano

One final, surprising application of flint makes use of the material's most underutilised quality – the pitched sound produced when struck by another flint or implement. The potential of this to produce music was explored by a French musician, Honoré Baudre, in the 1800s. He laboriously constructed a xylophone-like instrument with flint keys, which he played with two smaller hand-held flints. Known as the 'silex piano', this ingenious invention was said to use around 26 flints selected to produce a note of individual tone and clarity. Baudre toured with his invention, performing in France and England where the fine musical quality of this curious instrument was much praised and admired.

Honoré Baudre and his silex piano.

the late seventeenth century, the use of steel mills in mines was short-lived. Two large colliery firedamp explosions questioned their use and they fell out of favour. By the early nineteenth century they would be replaced by the safety lamp.

Chapter Three

Flint Architecture

Understanding a Flint Structure

Reading flintwork

'Reading' a flint wall by looking at the material used (type of flint and mortar), the style and the execution of the build can provide a lot of information. Pattern, style and type of flint are not always exclusive to all eras, regions and reasons; however, in general there are some clear indicators to look out for when trying to understand the narrative of a flint wall's build related to location, age and function.

The type, shape and colour of flint used might indicate geographical origin at a micro or macro level. Chalk or gravel pit? Cobble or field flint? Brown or blue cortex? For example, both gravel pit and sea cobbles have been exposed to millions of years of grinding and rubbing against each other, either by the sea and tidal movements or by glacial movement. Both tend to be more smooth, round and regular in shape. Chalk quarried flint tends to be more irregular in shape. However, repeating patterns and similarities can also occur in the irregular shapes and sizes, depending on age similarities and what biomechanical processes they have been through. Exposure to the elements, either in its raw state lying on the ground or built into a wall, can also change the appearance by creating patination. Popular terms for patination include toad belly and basket patination.

Mortar used may also indicate age and location of the build; for example, what type of lime used, any use of a pozzolan. Any shell content might indicate if it is river- or sea-dredged and therefore a coastal location. Mortar finishing and the

workmanship can sometimes indicate what time of year the work had originally been completed, or if the wall was built during a wet day or week. It should not make any difference, but you can normally tell these nuances.

The style of flint work, or a change in style, can indicate the age of the build as well as location. A good example of this is All Saints Church, Wheatacre, Norfolk. Not only can you see the window design change from the original fifteenth-century build to the nineteenth-century addition, there is also a clear historical change in flint style. Trends in flint styles have varied over the centuries.

Blue patination. Exposure to the elements, either in its raw state lying on the ground or built into a wall, can also change the appearance.

A flint detail showing an example of two different periods and types of build. All Saints Church, Wheatacre, Norfolk.

The amount the flint has been worked and tightness of the joints might indicate the status of the build in terms of location or patron. Closer inspection of a wall can sometimes indicate a narrative or tell you the history of the wall, intent, status, changes and adjustments to the area or purpose. Ground levels can change and heights of walls can change. This may be due to function, or simply to fashion. Often design and finishes are subject to not just ability but also available material.

The chancel wall of the Church of the Holy Trans-figuration, Great Walsingham, Norfolk is a good example of the style and material in conflict. The side facades are random snapped gravel pit flint with flush pointing. It is plumb and well laid. What is more intriguing is the front elevation with coursed snapped gravel pit flint (of the same source as the front facade) but with overlaid square pointing. First interpretation might assume that the mortar was left on by accident. However, it is surely too even and systematically squared-off for that to be the case. It might well be an aesthetic decision, however this is most unusual for both the period and the area.

It might well be that the original design was meant to be gauged squares, but either there was a lack in knowledge and ability to knap the flint, or the flint was too small to acquire any decent face. The latter is most likely of the two, as the flint has been cleanly struck to gain a 'fair face'; therefore, the skill was there.

Curious pointing finishing on the side elevation of Holy Transfiguration, Great Walsingham.

A random gravel pit flint on the front facade of Holy Transfiguration, Great Walsingham.

The execution of the work may indicate the ability of the tradesmen or number of people laying the wall. Similar to handwriting – although everyone may be using the same language – each individual has their own particular laying quirks and preferences. A good example of this is the visual difference between a left-handed or right-handed flint layer. This can sometimes just be isolated examples, but also of whole villages that have flint laid solely left handed by the resident flint layer. More visible in coursed work, right-handed work tends to be laid at an eleven o'clock tilt, whereas left-handed work tends to be laid at a one o'clock tilt. The cause of both is if the work is being laid left to right or right to left. This will be dictated by the placement of the perpendicular mortar bed and the angle of flint placement and compression. Although less obvious, you can also see these nuances in random work and gauged flush work.

Occasionally, there are laying styles and finishes that are hard to interpret and understand the original narrative of the build. A curious example of this is the vestry wall of St Peter's church, West Blatchington, Brighton and Hove, East Sussex (*see* page 30). Built in the late nineteenth century, the quality of workmanship, flint and mortar all appear to be from the same period. However, it is clearly evident that there is a style change from rough coursed gauged flush work to random flush work. With tight mortar joints and well-worked faces, the quality of workmanship remains and, in my opinion, has clearly been built by the same person. However, I have always assumed that the change in style was due to either a realisation of insufficient quantity of sizable and workable flint to

A left-handed coursed field flint with a 1 o'clock tilt.

A right-handed coursed field flint with an 11 o'clock tilt.

A flint detail showing a change in laying styles, St Peter's Church, West Blachington, Brighton and Hove.

make gauged work, or that they were taking too long and told to speed up. We will never have the answers, or know the reason why, but it is a good example of trying to understand the narrative of the build.

Raw material

It is clear that geology, functionality, changing historical industries and fashion have all influenced regional distinctiveness and the aesthetics of flint structures. Often the strengths and limitations of the raw material will define the structure or use. It is not just the regional geology of flint as a raw material, but also which exact strata (on a micro level). The time the raw material has been exposed to geological biomechanical influences such as 'freeze thaw' can have a huge influence and change the molecular structure of the flint. Therefore the origin and source can possibly predetermine, and definitely influence, the architectural options and outcome of the finish and style. Even as early as the Neanderthal period, people working or mining flint for tools and weapons realised the structural property differences between not only sources, but each individual geological flint strata bands. Grimes Graves in Norfolk is a perfect example of this, as there is clear evidence that the most mined and sought-after flint came from the floor bed. Due to the molecular structure of the flint

found at that level it was much more predictable to work with, meaning better outcomes, less time and waste. In terms of both current lithics and architectural raw material sourcing, nothing has changed. A knowledgeable flint worker will always be selective and careful in choosing their source of flint depending on the required outcome.

Geography and hierarchy

As a general rule, the change from function and functionality to aesthetics was an important influence. Although using the same material, the desire for looks and aesthetics outweighed the need for speed if protection was required. Evolution and a changing world have made their impact on architecture. The Industrial Revolution, the improvement of the transportation system and the changing needs of society have all had an impact.

As well as clear geological distinctiveness in flint architecture and the potential of the material, there are also other clear distinctions and influences. These may include geography, historical influences such as politics, religion, industry and economics, fashion, or maybe purely the desire to be different. Naturally, local vernacular material would normally be used, particularly before the improvement of the transportation system. There are numerous examples: the banded brick and flint work of Wiltshire and Hampshire, the brown random field flint work of the Chilterns, the use of galleting in West Sussex, the banded relief work of High Wycombe, and the miles and miles of coursed field flint boundary walls that divide the landscape of the South Downs.

There is also often a hierarchy of flint structures. A knowledgeable eye should be able to identify the status and importance of a wall (or at least the patron). Not only does flint work indicate skill, but also the quality of the material, and the time and money spent on the build. During certain periods of industrial growth there was huge prosperity for a select few industrialists and landowners who invested in a number of large country estates. Money was spent to showcase their wealth and emphasise how well

An example of the use of vernacular flint in a coastal location, Sheringham, Norfolk.

Coursed field flint boundary walls, Crowlink, East Sussex.

travelled and cultured they were. Many stunning buildings were built during this era. It was a time when resources, labour, materials and a good knowledge base were in abundance. On these grand estates a hierarchy of flint structures began to appear. The highest quality material would always be used on the main house ('the manor house'), followed by the estate church and rectory, then the main farmhouse, the workers cottages and finally the agricultural buildings, hay barns, threshing barns, livestock buildings and field boundary walls.

As well as these influences on the distinctiveness of flint styles, there also appear to be regional differences. For these there is no real explanation, except perhaps a result of the fashion of the time, personal preference of the patron or the imagination of the flint worker.

Historical Flint Use in Architecture

The first clear evidence of flint used in architecture was by the Romans. The Romans used vernacular materials and utilised natural resources throughout Europe. Flint in buildings and roads was no exception. These included fortifications, especially in coastal locations, palaces and buildings of status. Roman structures were in general crude buildings, with speed and function often being prioritised over aesthetics. The most common method would be *Opus incertum*: having a concrete, undressed chalk and rubble inner core, and flint as the external face. The flint laying style would either be random unworked flint (this was the Etruscan Way, which had a heavy Greek influence), or *Opus spicatum* ('spiked work'). We now know this style as herringbone or left hand, right hand work. It was allegedly inspired by the shape and pattern of the herring fish. The pattern was used throughout the Romans' European empire, as it utilised the maximum surface area of the flint. The Romans believed that it also aided strength. As a result, this pattern was used in roads as well as buildings. There are also a few occasions of *Opus mixtum* (flint with brick or tile banding), or examples of crude *Opus quadratum* (coursed flush work). It is unclear if the former was for structural or aesthetic purposes.

Not a lot of evidence of Roman flint buildings remains. St Albans Palace, Silchester city and walls, Portchester and Burgh Castle Fort are just a few examples that do. Many post-Roman buildings – especially fortifications – were built on the original Roman building footprint, but with little original evidence showing. However, an unusual one remaining is the Dover Pharos. It is reputedly the tallest Roman structure in Britain. It is a Roman lighthouse built on the edge of Dover cliffs to aid navigation across the channel from mainland Europe.

During the Anglo-Saxon period, flint was still used in architecture relatively infrequently – only on high-status builds, such as ecclesiastical buildings, priories and basic fortifications. Flint would still be used functionally and rarely aesthetically. Timber was still the preferred material of choice. The Saxons also had a few sea defence shore forts that were also adapted from Roman structures. But on the whole it was known that the Saxons were brave fighters and did not believe in stone fortifications. The Saxons liked fighting but did not see the importance of buildings in ruling. They were well into their use of wood structures, thatched roofs and clay-and-cob buildings. Reculver monastery on the North Kent coast is an example of the exception to the rule. Built in the seventh century, and adapted from a previous Roman site, it was built to provide better protection from Viking attacks.

It was the Normans who really introduced masonry and especially flint as the most common building material. French expertise, knowledge, skills and resources meant a lot more stone and buildings and use of vernacular material, such as flint. Coming from a Viking background of being hungry, ambitious and with a great desire to learn, they were resourceful and quick to transform their skills from shipbuilding to architectural buildings. These were taken from a big tradition of fortifications on the continent, and they brought that tradition with them. The Normans recognised the importance of strong defences and fortifications and defensive shore forts were their foothold in Britain. Pevensey castle is a typical example of this. Built on the former Roman site, they started building Pevensey almost immediately after invading. Due to the politics of the time and the influence of the Church, religion was also a big influence on Norman society. As well as fortifications, it was a great period of Romanesque ecclesiastical buildings. Many abbeys, churches and monasteries were built, including Lewes priory, Castle Acre priory and Thetford priory.

It was during the High and Middle Ages that flint architecture underwent a major transformation. Religion, politics and industry were the major contributors and influences to this change. The size and quality of buildings became symbols of power and influence. One of the largest and perhaps most unexpected influences on flint architecture was the wool industry; it was not so much the process of the wool industry but the wealth derived from it, hence the term 'wool church'. English wool was held in high regard and in turn sold

The Roman-built walls of Pevensey castle. Note the Norman additions in the distance.

The Dover Pharos. It is reputedly the tallest Roman structure in Britain, built on the edge of the Dover cliffs.

The remains of Castle Acre Priory, Castle Acre, Norfolk.

at a premium. At the end of the thirteenth century, the wool industry was bringing in huge revenue. Religion and politics of the time was also influential on how these wealthy proceeds were spent. This coincided with the influence of the church on communities and individuals. There was a heavy influence between the Catholic church and the current monarchy. There was a boom in ecclesiastical buildings that were purely funded by the huge gains made from the wool and woollen cloth. These were sometimes a new structure that replaced a previously modest rural build, a renovation or addition that would increase the grandeur or status of an existing building. With little indication of origin or background, this would explain why today these often grand and elaborate flint structures in often low-status agricultural and rural settings sometimes now look out of place. It not only coincided with the decorative and grand period of Gothic architecture, but enabled the rich communities, landowners, wool merchants and farmers to have a way to display their own wealth, status and faith, and in turn the belief that this may ensure a place in heaven. It was then that decorative flush work was introduced. Built in the early fourteenth century, St Ethelbert's Gate in Norwich is one of the earliest examples of high-status flush work. Built in 1320 and later restored in the nineteenth century, the rose

window on the west side of the gate shows elaborate designs of shaped flints placed within limestone ashlar. Not only did this transform the use of flint work on buildings, but matched the desire of the time to raise the status of a religious building.

Sometimes this would be a single individual, but mostly funding would be from a group of families or individuals within a community. Wool churches and 'wool towns' became common throughout southern counties of England, but the biggest influence of the

The Rose Window, St Ethelbert's Gate, Norwich. An early example of high-status flushwork.

industry seen today on flint architecture is in East Anglia. The close proximity of the renowned northern European commercial wool trading centres of places such as Bruges and Ghent, and the fashion for flush work, produced notable fifteenth-century 'wool' flint structures, including St Osyth priory's fifteenth-century gatehouse, and the perpendicular style (late gothic) churches of Holy Trinity church, Long Melford and the Guildhall, Norwich. It was not just on large-scale projects that the grandeur of flint was embraced; on a smaller scale, many parish churches used flush work in a much-diminished scale.

After the sixteenth-century Tudor period, the combination of the decline of the wool trade and the English Reformation saw the boom period in building new 'wool churches' come to an end. Flint would still be used on the building of cottages and farm buildings, but any new church works would be more minor, functional, and devoid of decorative grandeur. The demands and influence of the new monarchy of King

High-status flush work as a result of the wealth derived from the wool industry. Holy Trinity Church, Long Melford.

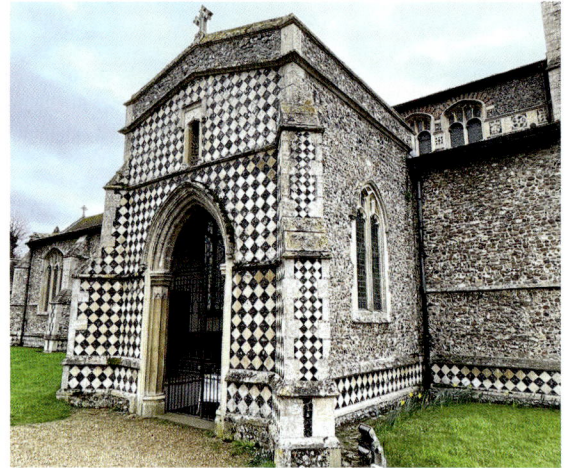

High-status, lozenge-shaped chequered work on a smaller scale, St Mary's, Walsham-le-Willows, Suffolk.

High-status, lozenge-shaped chequered work; a result of the wealth derived from the wool industry, the Guildhall, Norwich.

High-status, lozenge-shaped chequered work on a smaller scale, St Marys, Woolpit, Suffolk.

Henry VIII – that of moving away from the authority of the Pope and the Catholic church – and increased taxes to fund various wars and a lavish lifestyle took their toll. The destruction of the monasteries (1534) and the removal of land and any reminder of the power and wealth of the Catholic church in society played a fatal hand on ecclesiastical flint buildings. To an extent this was transferred to large Tudor palaces and estate houses. Flint would be used, but it was often mixed with brick or stone. An example of this was the Holbein Gate, sometimes known as the King's Gate. It was commissioned by Henry VIII and built between 1531 and 1532. The Holbein Gate was a triumphant gothic-style flint and Portland stone chequered gate that connected parts of the Tudor palace of Whitehall. Demolished in 1759, its materials are said to have been used on various buildings in Windsor Park.

This conspicuous display of personal wealth and status then continued throughout the seventeenth and eighteenth centuries. Some fine flint work can be found on grand estate houses throughout southern and eastern England. Many buildings constructed at the time were inspired by classical Greek and Italian Renaissance designs. This was a product of the eighteenth-century

Grand Tour. Also, built on estates at the time were numerous decorative and ornate arches, grottos, follies and even pigeon houses. Many existing structures were adapted or exaggerated to play to the romantic narrative of the time. A house facade within the grounds of Bury St Edmunds Cathedral, Suffolk was built around the tenth-century abbey walls and is in stark contrast to the gothic style of the cathedral.

It was a popular period for relief work. There are a number of fine examples of banded relief work in Buckinghamshire, particularly in the Amersham and West Wycombe area. Of note is the rather imposing West Wycombe mausoleum. Built in 1765 by Sir Francis Dashwood, this unusual hexagonal building sits on the hill above West Wycombe Park and village.

The Holbein Gate, or King's Gate. A gothic flint- and stone-chequered gate built by Henry VIII. An engraving from George Vertue's *Vetusta Monumenta* Vol.1, 1747.

An elaborate flint and brick detail from a pigeon house, Muntham Court, West Sussex.

It is very typical of the relief work of the area, using local field flint and Portland stone to create an elaborate array of Tuscan columns and arches.

It was not until the mid-nineteenth century during the period of the 'First Great Awakening' in America and the 'Anglican Revival' in England, when spirituality and religious devotion were revived and ecclesiastical building of any scale returned. The scale and status of this church-building period was intended to not only enhance the spiritual life of its congregation but to gain new followers. Add to this the expanding population and new churches were being built, or existing ones were being extended, restored or rebuilt. This period coincided with an abundance of resources, those of skilled labour and good access to high-quality raw material. Some very fine examples of high-quality gauged and decorative flush work were completed during this time. These include the chancel flush work of St Michael Coslany, Norwich (1883–1884), and the fine-gauged flush work of St Peter's and St Paul's church, Cromer, and Bury St Edmunds cathedral.

This period also coincided in part with the Industrial Revolution. The period from 1870 to 1890 saw the greatest increase in economic growth within such a short period than had ever previously happened. This was not only due to the increased wealth again

A house facade within the grounds of Bury St Edmunds Cathedral, Suffolk.

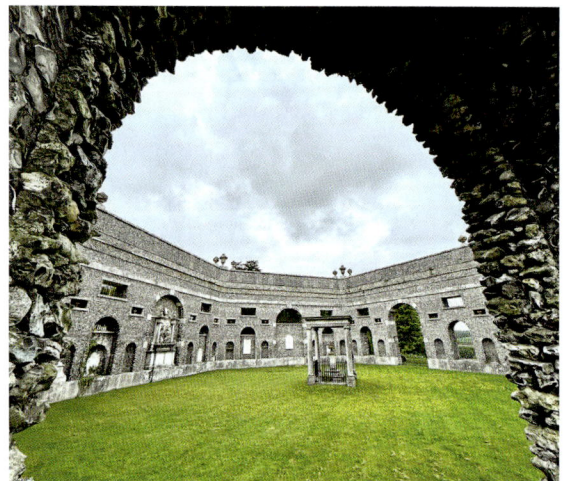

A flint arch detail, West Wycombe mausoleum. Built by Sir Francis Dashwood in 1765. West Wycombe, Buckinghamshire.

accrued from industry, and in particular the revival of the textile industry, but also from knowledge and skills gained from the gunflint industry, access to high-quality flint in its raw state and the improvement of the transportation system.

This period led to a widening of the gap between rich and poor. As factories and machines became more widespread, the wealthy factory owners and those who owned the means of production became richer, while the workers who were employed in these factories often remained poor. Flint manor houses, parish schools and community buildings would be built for the expanding population, but the boom in the house-building period for the workers in the mid- to late nineteenth century mainly used brick rather than flint. Flint walls would often be used to separate properties and define boundaries. However, as throughout history, the rich wanted to show how well they were doing and often the best way was to build with grandeur and often build with flint, but normally

A gauged flush work chancel detail. St Peter's and St Paul's Church, Cromer, Norfolk.

High-status flush work on the rebuilt tower, Holy Trinity Church, Long Melford.

High-status flush work on the rebuilt tower, Holy Trinity Church, Long Melford.

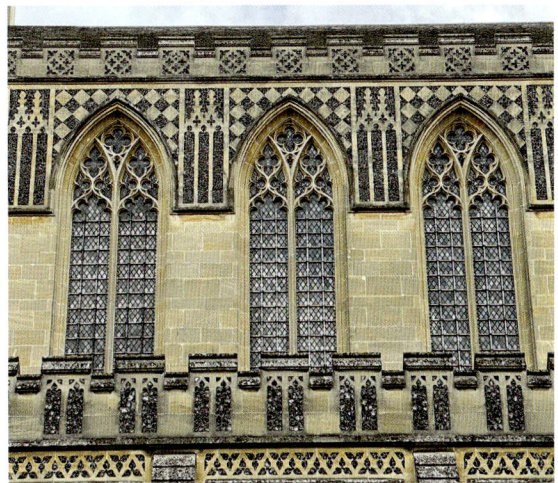

A gauged flush work detail, Bury St Edmunds Cathedral, Suffolk.

in combination with brick. The Tudor style-designed workhouse for Amersham Union of Parishes shows this combination well. Built in 1838 to provide work and shelter for those who had no means to support themselves, it is now part of Amersham hospital.

The growth of heavy industrial material also brought more new building materials, which included cast iron, steel and glass, with which architects and engineers rearranged the concept of function, size and form. More complex, elaborate and decorative designs were commissioned. As well as manor houses and religious buildings, the nineteenth century was also a period when flint was used in fortifications and defensive structures. These were mainly situated in coastal locations along the south and east coasts, and included Martello towers at the turn of the century, and more sophisticated mid-century Napoleonic fortifications.

In the early twentieth century, with interest in revival and nostalgia led by the Arts and Crafts movement, flint had a small revival. However, this soon ended in the 1930s, with the desire for 'modernism' in architecture. This heralded clean lines and form over function to escape the detailing and decoration of Victorian architecture. Render, brick and stone were suddenly all in fashion.

In Britain after World War II, pre-cast flint panels were manufactured on a grand scale. Due to the level of domestic destruction during the war, large-scale post-war rebuilding and redevelopments were necessary to re-house and re-build the cities' damaged infrastructures. There was a new building boom and, to cope with the current trends in design, the demand of this greater scale of manufacture required, and the shortage of skilled craftsmen, pre-cast panels were

A cobble and brick Napoleonic fortification, Shoreham-by-Sea, West Sussex.

A brick and flint facade at The Old Workhouse, now part of Amersham Hospital, Amersham, Buckinghamshire.

An Arts and Crafts flint and thatched house, Seaford, East Sussex.

Pre-cast flint panels used in Brandon, Suffolk.

Pre-cast flint panels used on the Catholic church of Saint Gregory, Eastbourne.

the perfect answer. It was a quick, simple and cheap form of construction. It was also a mass-production process that addressed demand and the development of both construction methods and building material technology.

However, it was not until the 1960s and 1970s that the demand for speed, together with the fashion for function and form, addressed the renewed awareness and respect of vernacular materials and styles. Flints of all shapes and sizes were suddenly used in pre-cast panels. In general, during the 1980s and 1990s flint was often only used for planning requirements rather than just aesthetics. People seemed scared of using flint. Whether this was due to lack of resources such as skilled labour and materials, lack of design awareness and potential or just safety in the familiarity of brick, it is impossible to say.

Over the last 30 years, flint has continued to be recognised as a vernacular material and used as a source of building material for new buildings. In 2004 the Wormsley Library was built. Situated on the Stokenchurch Estate, Buckinghamshire, this was one of the largest flint projects for a number of years. Fairly traditional in design with effective colour banding and relief work, it received little public attention. Following this build, flint continued to be used for a range of new builds for both commercial and domestic purposes.

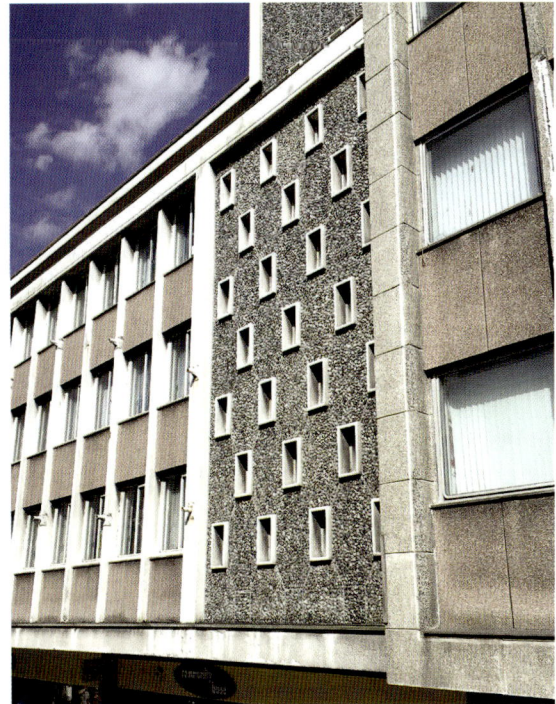

Decorative flint pre-cast panels used on Hanover House, Brighton and Hove.

Token flint use on a brick building facade, Fakenham, Norfolk.

A flint and brick detail, Holt, Norfolk.

The Wormsley Library, Stokenchurch Estate, Buckinghamshire.

However, the last ten years has seen a huge shift in popularity in how flint has been used. The 2014 Flint House designed by Skene Catling de La Peña for Lord Rothschild and built on the Waddesdon Estate, Buckinghamshire, rekindled interest in the material and its potential.

The flint work in this building fades from six different styles of flint throughout the elevation of the build. There is a transition from random coursing at the base to coursed at the top. The mortar colour also changes from black at the base to white at the top. To achieve the correct tonal range, all the flint was hand selected and came from six different quarries. The use of different size gallets from each flint source emphasised the texture but also enabled a smooth transition in colour and style.

The Flint House received a lot of media attention and won numerous awards, including the RIBA Manser Award for building of the year. Its unusual modern design certainly raised the profile of flint, and the potential of the material to many designers and architects. It was clear that not only the resources of skilled labour were available, but also the material. However, in my opinion the 2016 House 19 designed by Jestico and Whiles in Amersham, Buckinghamshire, had the biggest direct influence on domestic flint buildings and the industry as a whole. Having said that, House 19 was heavily influenced by the building of the Flint House, and may have not been built if the latter had not been built. House 19 won awards but, in my opinion, it was the accessibility of the project that appeared to many designers and self-builders.

The Flint House, Waddesdon Estate, Buckinghamshire. Designed by Skene Catling De La Peña, the building transitions from black to white and six different styles of flint.

A flint gallet detail. The Flint House, Waddesdon Estate, Buckinghamshire.

A flint detail. The Flint House, Waddesdon Estate, Buckinghamshire.

A flint grotto. The Flint House, Waddesdon Estate, Buckinghamshire.

The Flint House, as Described by the Architect

'Located on a seam of chalk that extends from the White Cliffs of Dover through to Norfolk on the east coast of Britain, the building is treated as landscape or geological extrusion. Flint is an ancient material related to jasper, obsidian and onyx; a hard, cryptocrystalline form of quartz found only in chalk, and in abundance on the surface of the ploughed fields surrounding the site. The architecture was generated from this neolithic material, the geology and the immediate ecosystem. The building is conceived materially as a flint landscape that has been carved away to become habitable. The flint was chosen for its rawness, to read like a Keifer painting of ash and lead, or the paint strokes of an Auerbach. The texture of the building answers the rough clods of ploughed earth in the surrounding fields. The flint is graded from a coarse, oily black, rusticated base to a refined fading out of smooth, matt chalk.'

Skene Catling de la Peña

House 19 designed by Jestico and Whiles. Random snapped flint and random snapped flint quoins.

House 19. Random snapped flint gallet detail.

Flint use in commercial buildings. Tesco superstore, Sheringham, Norfolk, designed by S + SA Architects.

Gainsborough House. Random snapped flint and brick detail. Designed by ZMMA.

It made random snapped flint and flint quoins very popular, especially in combination with either powder-coated zinc, corten steel or burnt larch. Since then there have been a plethora of random snapped flint projects all over the country, including the recent trend for internal flint feature walls.

It is not just random snapped projects and feature walls being built. There are other numerous reasons that have created interest and influenced architects, designers and self-builders to explore new flint combinations and designs. There is now a general new awareness of the extensive flint palette. I am glad to say that a greater awareness of the environment and the circular economy is also playing a role. The current potential of flint design is also being extended due to the use of modern technology and materials. An exciting example of this is the use of water-jet cutting to enable mechanical fixing on projects that would have been previously impossible.

The State of the Industry Resources Today

As with many trends and fashions within architecture, there can be an issue of the deficit between skilled labour and demand. It is clear that within some of the many recent flint builds – especially the domestic new builds – some are well executed and some less well executed. Nonetheless, whatever the quality of the build it has transformed the supply and skill base. Some chalk quarries that previously sold flint as a by-product to the agricultural lime industry have seen the financial opportunities and diverted their

Designed by Burrell Foley Fischer LLP, this flint roof uses a modern approach to flint design.

Random snapped flint domestic new build facade, designed by Colman Architects, Slindon, West Sussex.

The Hermitage, East Sussex, designed by Baker Brown Studio, uses random field flint in more modern form.

An example of an internal random flint feature wall.

Waterjet-cut flint enabling the mechanical fixing on flint in modern design.

A waterjet-cut flint lintel and flint ceiling, Depot, Lewes, East Sussex.

resources to selling flint as either a raw material or an upgraded processed/worked material. The accelerated demand of material and skilled labour has had a negative impact on the quality of builds. It has also increased the depressing use of flint blocks to manage demand. Yet it is not all doom and gloom: I can think of numerous small companies around the UK who are constructing high-quality work. The increased demand has also meant improved opportunities for people to gain apprenticeships and learn skills and has also seen clear evidence of a healthy number of young apprentices who clearly have potential and are keen to learn.

Architectural flint knapping is classified on 2021 The Heritage Crafts Associations Red List of Endangered Crafts as 'Endangered'. Those classified as 'endangered' are:

...those which currently have sufficient craftspeople to transmit the craft skills to the next generation, but for which there are serious concerns about their ongoing viability. This may include crafts with a shrinking market share, an ageing demographic or crafts with a declining number of practitioners (HCA).

I would agree with the first part of the classification, but not necessarily the second. I believe the flint industry is in a fairly healthy position. When I first started, there were no local firms that were solely flint specialists. There were bricklayers and general builders that would 'have a go'. Some were fairly competent; however, the use of materials was often incorrect. That was not necessarily due to habit, but more down to a lack of knowledge and availability of materials. There is now a plethora of firms up and down the country that specialise in flint, which can be found on the internet. This does not mean that they are all good: but it does show that there is a demand and an interest. I know of a number of competent skilled tradesmen that are of a high standard, but are still interested in learning, questioning current good practice (a healthy sign) in current processes and use of material, and retaining a high profile of flints. This does not mean we should be complacent. There should always be more encouragement and training for new practitioners interested in flint.

Industrial Historical Influences on Flint Use in Architecture

It has not just been the impact of major historical industrial movements that have influenced flint architecture. Specific industries have had their influence on architecture at a national and sometimes localised level. Below are a few examples of these influences.

The gunflint industry

In Suffolk and Norfolk in the eighteenth century, the expert skills of the flintlock industry heavily influenced the vernacular architecture of the area. However, this phenomenon was also very entwined with geology, in that the gunflint industry was very strong in certain locations due to the type and quality of the local flint. It was, however, not just the potential and predictability of the raw material, but also the increased knapping skill levels learned from producing flintlocks for the gunflint industry. These new skills were put to good use to influence the very elaborate and decorative flush work that

parts of Norfolk and Suffolk are famous for. It was not just the impact of the improved skills and better-quality material that changed some buildings. Of interest must also be the use of waste gunflint cores. Numerous houses in Brandon, Suffolk are faced using these unwanted nodules, which are all a result of the gunflint-making process. These are formed during the flaking process. This is when the outer edges of a large flint nodule are removed. Continual working of the outer edge will produce an inner core that becomes less workable and therefore for efficiency is discarded. In them you can clearly see around the outer edge of the core the shape of where flaking has occurred.

As well as gunflint core walls, Brandon also features a number of buildings made from another by-product of the gunflint industry – waste from the quartering process. This process assessed the quality of the flint and reduced it to a workable size. These waste fragments tended to be angular and irregular in size. Due to their varying size and shape they tended to be laid un-coursed and well interlocked. They were used on low-status buildings, often workers' cottages or outbuildings, and often used on the rear or side elevation with the more consistent flint cores generally used on the front facade.

Gunflint core waste used on the front of a house, Brandon, Suffolk.

Gunflint-quartering waste used on a house in Weeting, Norfolk.

Ironworks

Clinker is a by-product of the iron foundry process. Sometimes this clinker in lump form can look remarkably similar to knapped flint. In Lewes, East Sussex, there are numerous examples of walls being built in just clinker, but also in combination with flint. It is another good example of utilising the available material. This was a waste material being produced in excess – why not use it? The original foundry that supplied the clinker is no more, but there is evidence throughout the town of its influence. It was not just clinker that the foundry supplied to the local building trade. For some period of time the foundry would use sand moulds in the ironworks process. When the sand moulds were spent and unusable, sand would be available for use in the building trade. I only found this out by chance when working on a project in the local area. The project involved numerous repointing and rebuilding items to a local church. Despite the church originally being built in the twelfth century, as with many structures there have been numerous alterations and repairs made over the years. On a visual inspection I had made the assumption that the black particles in the mortar were pozzalans (*see* the section on mortar). However, when undertaking a proper, more detailed mortar analysis, I came to realise that these particles were something else. It was only by enquiring of some local knowledge that I was informed that the black particles in question were clinker and that the sand used for certain repairs was called 'Phoenix sand'. This was a name taken from the local iron foundry that supplied the sand from the spent casting moulds.

The fishing industry

Another example of industrial influence on coastal flint is the cobble flint style of 'fish scale'. This is when the base of the cobble is removed and worked into a concave shape. This means that the cobble can be laid at a tighter fit than if it is left unworked. Although there is no clear historical evidence to prove it, it has often been assumed that the name and appearance

Clinker used in a boundary wall with flint, Lewes, East Sussex.

'Fish scale' flint style, often found in coastal locations and possibly inspired by the local fishing industry.

Coursed snapped and unsnapped cobbles, with flat gallet placing, Glynde, East Sussex.

reflected the fishing industry and scales on a fish. One can clearly see similarities. Whether this was the original inspiration we will never know, however it was a very effective laying style that reduced the use of mortar. Sometimes this was related to status, sometimes a cost benefit. In a period when cobbles and labour were cheap, lime mortar was not.

Agriculture and ceramics

Glynde Estate is another example of quirks. Glynde is an inland Sussex downland village situated 6 miles (10km) from the sea. Since 1086, as well as general agriculture, it was known for its production of salt. Its flint vernacular should be field flint; most of the houses and agricultural buildings in the village are made of field flint. However, the manor house, Glynde Place and the Chapel both contain the extensive use of sea cobbles in their design. It transpires that during the period that the chapel was built (1765) and the major alterations to Glynde Place, the estate farm manager was John Ellman (1753–1832), famed for his breeding of the Southdown sheep. As well as a sheep farmer and progressive employer, Ellman was influential and well connected. He saw the importance of trade and was instrumental in not just the improvement of the navigation of the local river and therefore access to the village but played an important role in the redevelopment of Newhaven harbour, the nearest coastal trading point. Newhaven was also a well-known centre and trading point for cobble supply to the ceramic industry, which was in full flow at the time. Why is it relevant? Well, either Ellman had easy or free access to sea cobbles at Newhaven, or he saw an opportunity and the potential of the sea cobbles being used as ballast in the flat-bottomed sailing barges that would dock and trade in Glynde village. Barges would sail upriver from the coast with sea cobbles as ballast, shed the now unwanted cargo and load up with grain, wool or other goods.

Chapter Four

Flint Styles

Field Flint

Field flint is sometimes known as unworked flint, found flint, whole flint or dry flint. It is probably the most used flint in building. As its name suggests, it is taken directly from the field and normally used unworked. It is often humorously referred to as 'the one crop that never fails', because it has for centuries been the bane of a farmer's life. Field flint can reduce crop yield and damage farm machinery. In World War II, land girls were employed to walk the fields of southern England to remove flint. Through seasonal erosion, field flints work their way up through the soil and end up on the surface of the ground. Field flint picking used to be part of the agricultural calendar. During the winter months, after ploughing but before sowing and rolling, agricultural workers would walk the fields removing the flint. This would have a dual purpose: to clear the fields, but also to create a stockpile of building material for the year ahead. All over southern England there are villages entirely constructed of flint.

Field flint can vary in both size and colour. Depending on the region or the mineral content of the soil, the flint colour can change. This can impact both the flint rind and inner material. Colours can vary from shades of whites and greys (Sussex and Kent), to blues (Hampshire and Wiltshire), browns and yellows (the Chilterns, Hertfordshire and East Anglia). This is by no means exclusive to each area. Colour variation can be found within a surprisingly close proximity. As they sit on the top of the soil, time and exposure will often bleach flints to shades of white. The name 'dry flint' comes from the brittle and unpredictable nature of the material. Thermal fractures occur as the material moves up to the surface, making it very unworkable. As a result, field flint is often considered as a low-status flint used in rural environments. Styles and patterns of field flint designs can vary depending on region, status or period of construction. As a general rule, the southern counties of England tend to lay field flint course, whilst the home counties tend to lay it at random.

Mortar joints can differ from recessed, flush and decorative pointing. Depending on coursed or random coursing, flint sizes can vary enormously. However, if coursed, some grading will occur to retain the gauge. Field flint fragments are a popular use of low status buildings in rural and agricultural settings.

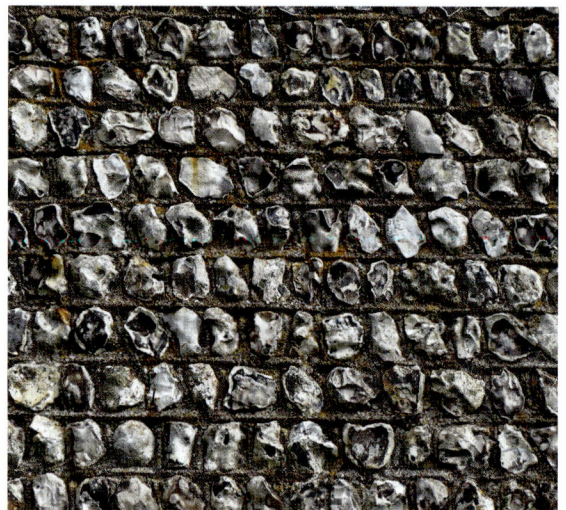

Coursed field flint with weather struck pointing.

Coursed field flint with flush raked-back pointing.

Random field flint with recessed pointing.

Coursed field flint with weather struck undercut pointing.

Random field plate flint with weather struck pointing.

Knapped or Snapped Work

Knapped or snapped work – sometimes also known as select work – is when a flint has been split open to expose the core of the stone. As a predictable source material is required for the shaping process, snapped work tends to be made from quarried flint, from both chalk and gravel pits. It is laid with the flat face facing out, and can be laid with or without the cortex showing. Snapped work tends to either have no or little excessive working or squaring off

from the flint; just sufficient enough shaping to fit well next to existing flint and reduce the size of the mortar joint. Joint size can vary depending on status and ability of workmanship. Laying patterns vary from coursed to random. There are early examples of fourteenth-century coursed snapped work; however, it was not until the eighteenth and nineteenth centuries that snapped work became very popular, particularly random snapped work with dark mortar. This popularity in style appeared to coincide with the increased availability of higher-quality quarry

was previously sought after as it signified high quality. However, this is not always the rule, there are some very fine examples of other colours, or work laid with an alternating colour contrast. These include brown or chocolate tones on St Peter's and St Paul's, Cromer in Norfolk and the popular chequered board gauged flush work on the north coast of France.

An additional factor to note is that the nature of working gauged flush work means the depth of the flint will often be quite narrow. For this reason, the structure is often reliant on the inner substrate for strength. Therefore, it is particularly important to address any water ingress at the earliest opportunity.

Folkington Manor in East Sussex is an example of unusual narrow-sized gauged flush work. On this build there is a much higher percentage of flint per square metre. This could have been an aesthetic choice; however, I am not convinced small flints and additional knapping would have been selected out of choice. After visiting the local chalk pit quarry (a likely source of flint for the build), I am convinced that this decision was a result of the size of flint available.

Chequered board gauged flush work, typical of Normandy, France.

Chocolate-coloured gauged flush work, St Peter's and St Paul's, Cromer, Norfolk.

Narrow gauged flush work, Folkington Manor, East Sussex.

A tight mortar joint on coursed gauged flush work.

Cobble Work

Depending on style and regional variations, cobble work can also be known as sea flint, rolled flint, whole stone, duck egg or fish scale. There are numerous other variations of this type of flint work. Matching the vernacular, they are typically found in coastal locations. They have an effective texture and are often used for relief work. Colour variations depend on local regional geological differences. As a general rule, there is a big difference between the brown and yellow tones of the gravel pit and beach-sourced flint of East Anglia, and the grey- and blue-toned cobbles on both sides of the English Channel.

Subject to location and status, laying patterns vary from random and coursed, and from snapped to un-worked. Coursed work normally tends to be gauged, unworked and often used in conjunction with bird's beak pointing. There are also numerous examples in coastal locations within very close proximity to the sea, of pitched or tarred coursed cobble work. It could be assumed that this is due to the combination of aesthetics (darker flints are often fashionable and considered 'higher status'), but also for practical reasons. The high atmospheric salt content in coastal locations reacts with lime mortar, causing accelerated deterioration. Over the years we have certainly had to repair and stabilise a fair amount of such walls. As this solution is normally only found in coastal locations, it might be linked to the availability and knowledge of tar in the local boat- and ship-building industries. It would certainly match the early to middle-nineteenth century time period of the heavy use of pine tar and coal tar.

There are numerous regional and coastal variations on cobble work – parts of east Devon are known for its 'Penny Bun' work. These are large rounded cobbles; I have been unable to find the precise reason and origin of the name. Perhaps it derives from the penny bun loaf, a historically common sized loaf. I have also been told that there is a link with the Penny Bun mushroom. They both look similar and this mushroom can be found locally in abundance at certain times of year. Whichever the origin, this size and shape of cobble and name are both very specific to the area.

Another example of regional variation of cobble work is 'duck egg'. This highly effective and decorative finish can be found in numerous parts of the country, but is particularly popular on the north Norfolk coast. As the name suggests, duck egg cobble work normally

Tarred course cobble work, often found in southern coastal locations.

Random snapped cobble flint with flush pointing.

Duck egg flint work, the Old Infant School, Burnham Market, Norfolk.

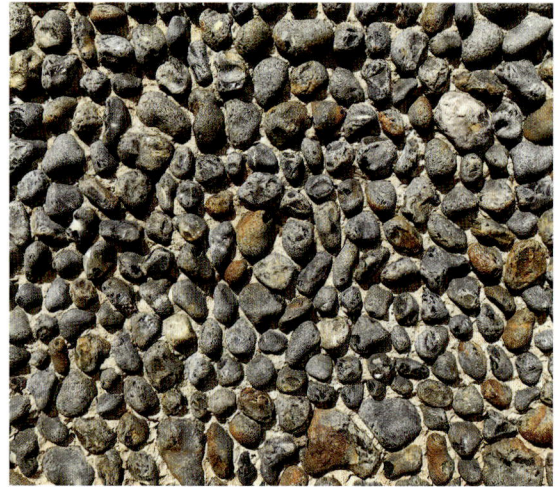

Random cobble work, coastal location, north Norfolk.

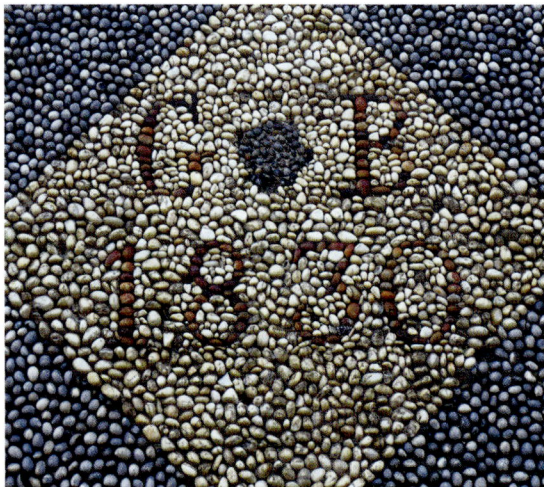

Decorative contrasting colours in duck egg flint in the village school, Overstrand, Norfolk.

Coursed cobble work with weather struck pointing, Rottingdean, East Sussex.

uses well-graded flints that are never larger than an egg. As well as being size graded they also tend to be colour graded. They are either all the same tone and colour or chosen for decorative contrast. Good examples of normal duck egg work include the Old Infant School in Burnham Market, Norfolk. Another good example of using decorative contrasting colours in duck egg flint is the village school at Overstrand, Norfolk.

Galleting

Galleting – sometimes known as garneting or garreting – is a combination of thin flint flakes, or pieces of stone or brick, that are pushed into the mortar during the building process. Galleting can be very effective aesthetically, but is one of the most time consuming and labour intensive of all the laying styles. There are numerous styles and nuances of galleting or flint and galleting combination in use. Styles include: density, angle and depth, white or black, or use with field or snapped flint, use with coursed work or random work. The density and movement of the gallets can markedly change the appearance of the wall. Most gallets will be pushed vertically or on the diagonal. This will aid the flow of water off the wall and limit any trapping or water ingress. Gallets are normally overlapped and work around the shape of the flint. This helps provide movement and flow. I was taught that successful galleting must look like flowing water or a shoal of fish. Butt jointing or pushing in the gallets too vertical and horizontal can make the overall effect very static. There is some debate as to whether the purpose of galleting is at all functional. It is clear that historically gallets were often placed to stabilise the flint and enable more height gain in a day's work. However, they do also support the joint

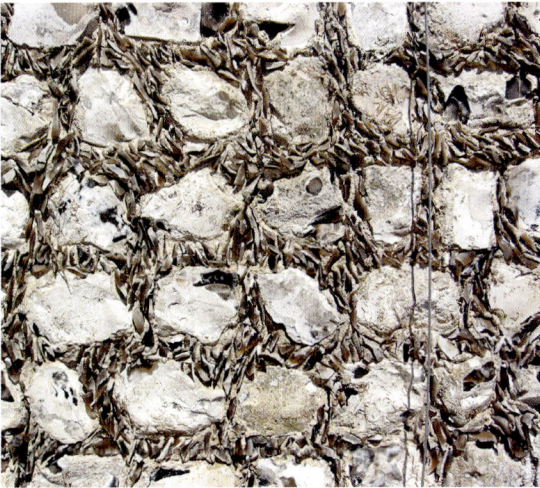

Coursed field flint with field flint gallets, West Dean House, West Dean, West Sussex.

Random gravel-pit flint with gravel-pit flint gallets, The Flint House, Waddesdon.

Coursed field flint with gallets, Chichester, West Sussex.

Random gravel-pit flint with oyster shell gallets.

Galleted front facade of house, East Runton, Norfolk.

Carstone chip galleting with coursed work, Stoke Ferry, Norfolk.

by protecting the mortar from weathering. It is also evident that sometimes the placement of flakes is an afterthought to fill in a large mortar gap, whereas some are intentionally laid in a regular or dense decorative pattern. While both of these are true, due to the time-consuming nature of making and pushing in the gallets, it is my opinion that gallets have been used purely for aesthetic gain.

The origins of galleting are not clear. There is some clear historical evidence during the Norman period of pieces of flint and other material used as packing and wedges. There is also some suggestion of flakes being used as a key for sneck harling (rough render). They would provide much-needed additional adhesion for rough lime render to the otherwise smooth flint surfaces. However, by the fifteenth century the quantity and density of some of the galleting work appear to only be for aesthetic purposes. There are examples in the late eighteenth and early nineteenth centuries of gallets being used as a deterrent in buildings (for example, prisons and custodial buildings). Dense sharp flint flakes are very unappealing, if not impossible to climb up. This period would likely have also been influenced by gunflint making and access to quarry flint during the boom-building period.

The use of flint flakes for gallets is mainly found in the southern counties of the UK. The use of galleting in West Sussex, particularly around the Chichester area, is very prevalent. Of particular note are the Goodwood and West Dean estates. Both were built in the eighteenth century and are remarkable for their skilfully worked gallets, flint quoins and window reveals.

On the north Norfolk coast are the villages of East Runton and Overstrand. Here, and in some nearby villages, are some remarkable buildings with a total facade made of gallets. These are quite unusual, as most appear to be built by the same hand, or hands.

Banded

Banded flint is when horizontal flint bands (normally coursed) are interspersed with another material. This may be brick, stone or a different style of flint. This style of flint can be found all over the UK, but is particularly popular in Wiltshire and Dorset. Built from the fourteenth century onwards, there are plenty of examples of vernacular buildings using flint and stone throughout the UK, and particularly in parts of East Anglia. Depending on the period of construction and

Banded brick and flint, typical of Wiltshire and Dorset.

Stone and snapped quarry-flint banding, Wiltshire.

Horizontal alternating colour banding, Buckinghamshire.

A detail of stone and snapped quarry-flint banding, Wiltshire.

Chess board and brick banding, Normandy, France.

Composite banding, Normandy, northern France.

status, the majority of examples are coursed field flint with double or triple courses of bricks. It is hard to know why the combination has been used and not just one style or the other. It may have been for a number of reasons: to raise the status of the flint work, or out of pure convenience, as bricks are so easy to use and there are often more bricklayers available. Depending on the availability of the bricks, it may have also been a decision based on cost effectiveness. I can also think of some examples where a brick band shoestring course (single course) has been laid to provide structural strength, or level and straighten up the flint work. That certainly would work if the person laying the wall lacked confidence in what they were doing. However, this is not always the case, and the choice of using banding can be purely aesthetic and very effective.

Chequered Work

Chequered work is often known as chequerboard or chessboard work. It is a check work pattern. It is created by an amalgamation of two materials in alternating

A galleted flint and stone chequered board, East Sussex.

Stone and flush work chequered work, Norfolk.

Brick and coursed snapped flint chequered work, Suffolk.

Chalk and snapped chalk-quarry flint chequered work, Wiltshire.

squares. Materials vary from stone, chalk, and brick, to different flint styles or contrasting flint colours. There are early fourteenth-century examples, but it became very popular in the fifteenth century. It is possible to find both tight chequered work (where a pattern is followed very precisely) and loose chequered work (where a pattern can be followed, but is less rigid and regular). There are some very good examples of chalk and flint chequered work in Hampshire and Wiltshire. This would typically be using black snapped flint to achieve a direct colour contrast to the white chalk. Chequered work may have originated for a number of reasons; flint was cheap, therefore using both may have saved costs, both structurally and aesthetically.

Relief Work

Relief work is when a flint is laid in combination with another material or another style of flint. It is a very effective way of using the maximum potential of the material in terms of texture and form. Patterns can vary, but often the use of random or interlocking laying style in contrast with flat brick

Gauged flush work and sandstone chequered work, Woolpit, Suffolk.

Relief work, Old Amersham, Buckinghamshire.

Composite chequered work, Basham, Suffolk.

Relief work, West Wycombe Park, West Wycombe.

or stone appears to be the most successful. It is definitely one of the more creative flint styles. The selection of the flint is crucial. It is impossible to hand select random-shaped flints from particular quarries. Hand selecting provides maximum opportunity to interlock the flints and show a more exaggerated contrast to the flat face of flush work, stonework or brick.

Relief work can be found in various places throughout the UK. A variation of relief work is 'folly' flint, which in modern times is sometimes known as 'baked bean' flint. As with relief work, the influence of the eighteenth-century Grand Tour is abundantly clear. A fine example of folly work is the classically designed Triumphal Arch on the Holkham Hall estate, Norfolk. Built in 1757 and designed by William Kent, the arch was built to extend the south driveway and was allegedly inspired by the triumphal arches of ancient Rome. It was built to impress. It is well executed in contrasting styles of

Triumphal Arch, Holkham Hall estate, Norfolk.

Relief work, Eartham folly, Eartham, West Sussex.

Triumphal Arch detail, Holkham Hall estate, Norfolk.

An Eartham folly detail, Eartham, West Sussex.

tightly packed random quarried flint, in contrast to flat ashlar stonework. Once the home of the estate farm labourers and shepherds, the gateway is now a holiday rental.

Decorative Work

Decorative work is many people's favourite style of flint work. There are no set rules with this style. As with flint relief work, the use of colours and textures of different materials or styles of flint can create an effective contrast. It is often the most creative and imaginative of all the flint work styles. Around the UK there are numerous examples of individual patterns that are an amalgamation of different materials. Designs can be the creative inspiration of the workers laying the flint, or the patrons commissioning the work. The most popular materials for decorative work are brick and flint. Although they use the same

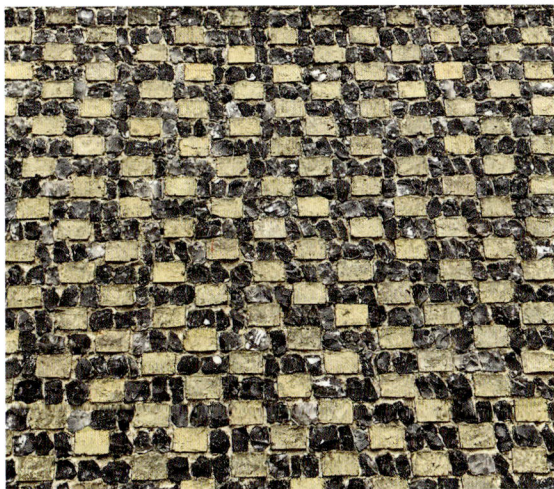

Decorative flint and brickwork, Norfolk.

Decorative random snapped chalk-quarry flint and stone blocks, Suffolk.

Decorative flint and brickwork, Hardwick, Suffolk.

Decorative snapped chalk-quarry flint, chert and boxstones, Suffolk.

Decorative squared gravel-pit flint and stone, Suffolk.

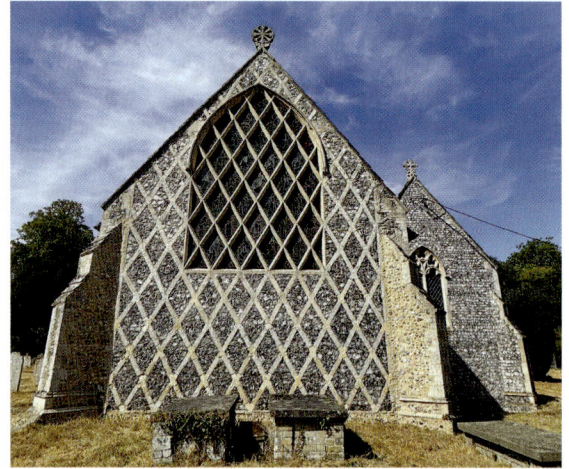

Diaper work on the east wall at Holy Trinity Church, Barsham, Suffolk.

materials as bungaroosh work, the deliberate decorative patterns of alternating patterns and combinations raises the status of the style and the work. There are examples from as early as the mid-fourteenth century; however, probably its most playful form is from the eighteenth century onwards. This work is particularly interesting as it was often used on low-status buildings such as small domestic cottages and buildings. There are numerous examples of individual motifs and patterns that are one-offs. These might include an individual decorative pattern, dates and initials. The style can be local to an area, a village, or a specific house. It may also often be down to the individual flint layer. Certain areas of Norfolk and Suffolk appear particularly playful, with materials and pattern use. It is hard to know if this is the 'Wool Church' influence, and the need to be different. Also, with some examples it is hard to know if the patterns have been guided by the materials available, or if materials have been acquired for particular patterns. Regardless, they are clearly very effective.

Diaper Work

Diaper work or lattice style is when two styles of flint (commonly flint and brick) are laid in an equal diagonal pattern, creating regular diamond or lozenge shapes. There are early examples of fifteenth-century diaper work, though it became most fashionable in Tudor and Elizabethan periods. This may be due to an increase in the use of bricks or the influence of wool trading routes and the Belgium and Flemish use of brick diaper work. There are no set criteria on the materials that are used. Although mostly combinations of brick and field flint or used, there are also some good examples of brick and cobbles or brick and gauged flush work. When the flint being used is black or grey, it is hard not to compare it to the use of glazed brick headers. An unusual but effective example of latticework is the flint and stone latticework to the window and wall of the east wall at Holy Trinity Church, Barsham in Suffolk. Most likely built in the late fifteenth century, it is a particularly good example of effective material use and design. Built in an area that was heavily involved with the wool and cloth trade, it is hard not to believe that this may have been influenced by the wool trading routes of the continent. However, there is also some conjecture about the design similarities between the latticework and the local landowners and patron Lord Etchingham's heraldic shield.

Detail of the east wall at Holy Trinity Church, Barsham, Suffolk.

Brick and random snapped flint diaper work, Methwold, Norfolk.

Brick and random field flint diaper work, Norfolk.

Snapped chalk flint and brick, Newmarket, Suffolk.

Brick and snapped chalk flint diaper work, Oxburgh, Norfolk.

Glazed headers and coursed snapped chalk quarry flint, Folkington, East Sussex.

Snail's Creep

Snail's creep became popular as a decorative finish in the mid- to late nineteenth century. It is normally used in conjunction with brickwork. Plate flints or flints laid flat were also popular. This would utilise maximum flint shape without showing any of the flint core. There are two ways to achieve a snail's creep effect: the traditional way of adding on the 'creep' in a finer mortar after the laying is complete (similar to brickwork tuck pointing), or creating the creep during the building process using the existing mortar. Unfortunately there are now many examples where over time the adhesion of the 'creep' has failed and has come away from the mortar joint. This may be down to poor execution in the original build, but it is clear it is a vulnerable joint application.

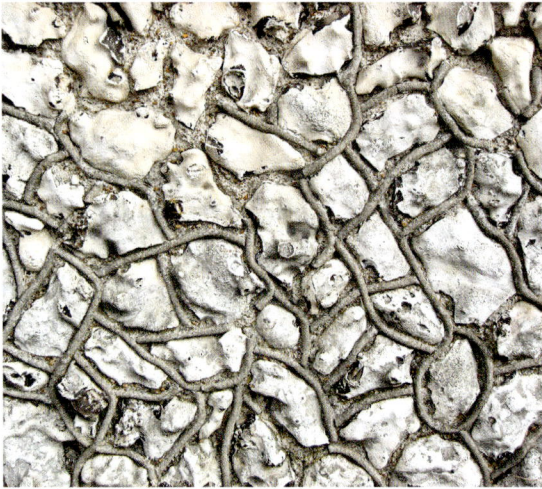

Snail's creep with missing 'creep'.

Snails's creep pointing detail.

Snail's creep pointing with chalk quarry flint.

In situ snail's creep red mortar pointing with field flint.

Decrative Flush work detail

Decorative flush work inscription, Stratford St Mary, Suffolk.

Decorative flush work, All Saints church, Hawstead, Suffolk.

Decorative flush work, St Mary of the Assumption church, Ufford.

Decorative Flush Work

Decorative flush work takes the form of architectural flint patterns and motifs that are normally found at the base of church buttresses, plinths and parapet walls. These are mainly found on vernacular buildings in East Anglia. They are very thin flint shapes that are worked and sit almost like veneer within the stone. Post-fifteenth century, late Gothic (perpendicular architectural style) decorative flush work was regularly used to update, or raise the status of a church by incorporating the style in new porches or additional annexes. It showed skill and status. Dark flint was often easier to work and was used to highlight the shapes and create contrast to the stone. Decorative flush work can reflect the architectural features of a building. They can be very elaborate and can come in many forms. These can be symbolic emblems, religious motifs, inscriptions and lettering, initials or heraldic symbols of the patron, or purely decorative in the form of repeated patterns, chequered work, trefoils, candles and geometric abstractions.

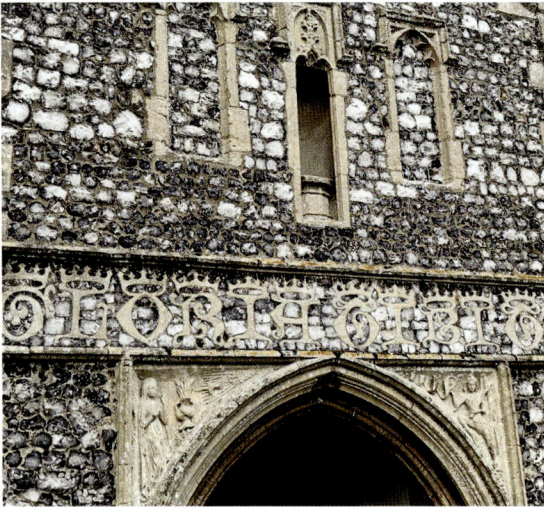

Flush work inscription, All Saints church, East Tuddenham, Norfolk.

Decorative flush work, St John's church, Elmswell, Suffolk.

Rubblework

Rubblework is usually known as 'bungaroosh' work. It is often considered 'low status' as it is a composite of any material available. Depending on the local area or the period of construction, this will be whatever material is available. Rubblework is found throughout the UK. In East Anglia there are many examples of churches, houses and garden walls with a mixture of flint, chert, chalk and brick. In the southern counties of the UK, there are plenty of examples of brick and flint combinations. Bungaroosh work can vary from completely random to highly decorative pattern combinations. In the Victorian boom-building period of terrace houses, no material went to waste. There is an area of Brighton, East Sussex, where rows and rows of terraced houses were built during this period. As the houses were being built, tradesmen would move from one house to another completing the same building tasks on each house. Often on completion of the houses, the garden boundary walls would be built using any leftover house building material waste. In this case it would be flint dug up from the footings and any broken half bricks that were

discarded during the build process. There is one street whose footings are perfectly lined up with a flint start band in the ground. Excess flint was also used on the internal dividing walls to the houses. This street has the unwanted nickname of 'picture hanger's hell'! Anyone nailing or screwing a fixing into the wall finds it almost impossible to line up two fixings without hitting a flint.

Coursed decorative chalk and brick pattern.

Coursed brick and flint rubble wall.

Coursed decorative brick and snapped flint pattern.

Rough coursed brick and snapped flint.

A random wall with flint, sandstone, chert and waste flint.

Quoins

There are numerous different styles of quoins. This style is very subject to the period and the status of a build. Crude field flint quoins were used in many Norman structures, including priories and fortifications. In the twelfth century, the focus was on functionality rather than aesthetics. Field quoins tended to be over-sized with very little workings. They were also often used for door-jams and window arches. As a result of the medieval emphasis on limestone, there was a reduction in the use of quoins, particularly Caen stone imported from northern France. In addition, the mass production of bricks reduced the use of flint quoins. Why work a stubborn material like flint to make quoins when bricks serve that purpose very easily? It was not until the early eighteenth century when flint had become fashionable again, and resources of labour, materials and skills were in abundance, that flint was again readily used for quoins. These were on high-status buildings such as grand stately homes or ecclesiastical buildings, and also many estate buildings. During this period there are many examples of quality random snapped and gauged quoins. Random snapped quoins are when quoins have been worked to a 90-degree angle, but the remaining part of the flint is unworked and random in shape. Gauged quoins are also part of gauged flush work. They are worked to a 90-degree angle, but also then worked on all edges to align up with gauged flush work or ashlar work.

At West Dean House, West Sussex (built 1622 but remodelled in 1804) there are some most remarkable convex and concave window quoins. These really do show skill, patience

Concave window quoin reveals, West Dean House, West Sussex.

Flint quoins on The Flint House, Waddesdon, Buckinghamshire.

Random field flint quoins, West Dean House, West Dean, West Sussex.

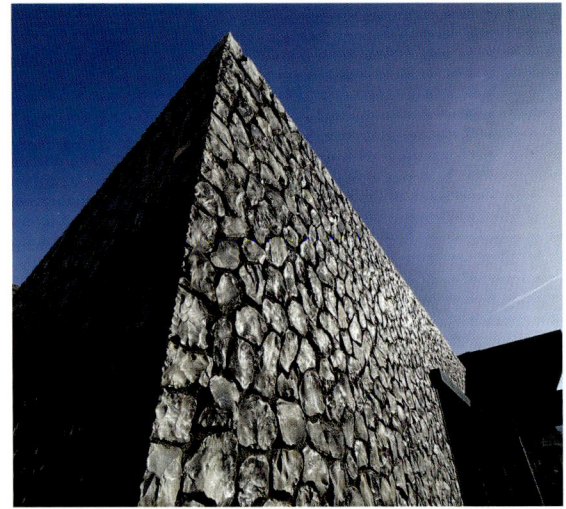

79-degree angled random snapped quoins, Depot, Lewes, East Sussex.

Window reveal detail, West Dean House, West Dean, West Sussex.

and an abundance of high-quality raw material. Although some of the flint is respectable enough of the qualities required to produce such work and would come from a local source (there is a quarry still open that matches the colour and qualities), from the state of the work it is questionable if the skill level was there. It is therefore no surprise that records show worked flint for West Dean and the nearby Goodwood Estate were packed in straw and shipped down in barrels from Norfolk, a region that at the time clearly had the skills and raw materials required.

Since 2013 and the widely publicised Flint House designed by Skene Catling De La Peña, flint quoins have made a resurgence. In particular, random snapped quoins have become very popular. The quality of these quoins does vary considerably; however, it does show that flint quoins (both random and gauged) are achievable and accessible.

Floors

The Romans were the first to use flint for floors and roads in England on a large scale, and it is understandable why they chose this material. Its abundance, regional availability and construction properties made it ideal. Flint could be found on the surface over large areas of southern England, reducing time mining or transporting other alternatives. The Romans' construction method was simple: they would generally lay dry roads using large nodules at the base, followed by layers of smaller flint and chalk particles, and top with gravel or fine particles. Their roads would always be cambered and contained by large flints or stones as edge or curb stones. Flint was an ideal material due to the durability and ability to withstand heavy traffic, and because its irregular shapes created a structure that interlocked and held in place well.

locations nearby are some remarkable buildings with complete facades made of gallets. These are quite remarkable and must have taken some time not just to build, but to make the gallets. If you look at the depth, the gallets are pushed into the mortar and are angled and overlapped. I believe them to be built by the same hand. I am not aware of any other buildings in the country that are solely laid in gallets. Putting the aesthetics aside, I do think that these facades with so many gallets are vulnerable to weather and losing gallets. Having said that, most of the facade of The Flint House in Overstrand seems to be original.

2. **Neufchâtel-en-Bray** (Normandy, France) – In the Normandy town of Neufchatel-en-Bray, there are a few buildings modest in size but certainly not lacking in creativity and craftsmanship. They are all quite localised within the town, and looking at both the knapping and the laying I assume they were completed by one person or a team of flint masons and layers. France is no exception to Britain, with a regional flint distinctiveness as a result of both historical trends and geographical reasons. Across the Normandy coast, the style of chequered work with alternating dark and light flints is present, specifically in western areas of Normandy, where the chequered work with raised pointing is present. There are many examples of raised pointing on the front faces of houses. Also popular is beehive pointing. In my personal opinion, these walls might be well executed, but the point really does take away from the shape and beauty of the flint. Again, most of these appear to be of the same period and likely laid by the same hand or hands.

3. **Whalebone House** (Cley-on-Sea, Norfolk) – A quite exceptional eighteenth-century building on the north Norfolk coast built of flint

A galleted house facade, Overstrand, Norfolk.

A galleted house detail,
Overstrand, Norfolk.

A galleted house,
Overstrand, Norfolk.

and bone. The skill and creativity in using flint and bone is quite astonishing – some may say, rather macabre – but I quite like it. As well as the relief work of flint and bone on the lower sections of the front facade, of particular note is the over-sailing cornice of vertebrae detail at the top. I am unaware of any other buildings in the British Isles like it. There are other floors or buildings with bones, but as far as I am aware, none built with such creativity and in such detail. It is very precise in execution and constructed with great understanding of the shape and form of each bone. Although it is rather romantically known as 'the Whale Bone Building', I have had the bones visually analysed by a vet and in fact there are no whalebones – just various bones from sheep, cows, pigs and horses. Although the building is now used as a house, it may be no coincidence that it was once a butcher's shop.

Creative lozenge and gauged flush work, Normandy, France.

Raised pointing, Le Tréport, Normandy, France.

Beehive pointing, Normandy, France.

Taken in Normandy, this photograph shows a flint and brick palette selection on the front facade of a house. I am not sure if it was the intention, but it very much looks like a sample board of flint work – a selection of styles created for the client or designer to choose from when the main build commences.

Bone cornice detail, Cley-on-Sea, Norfolk.

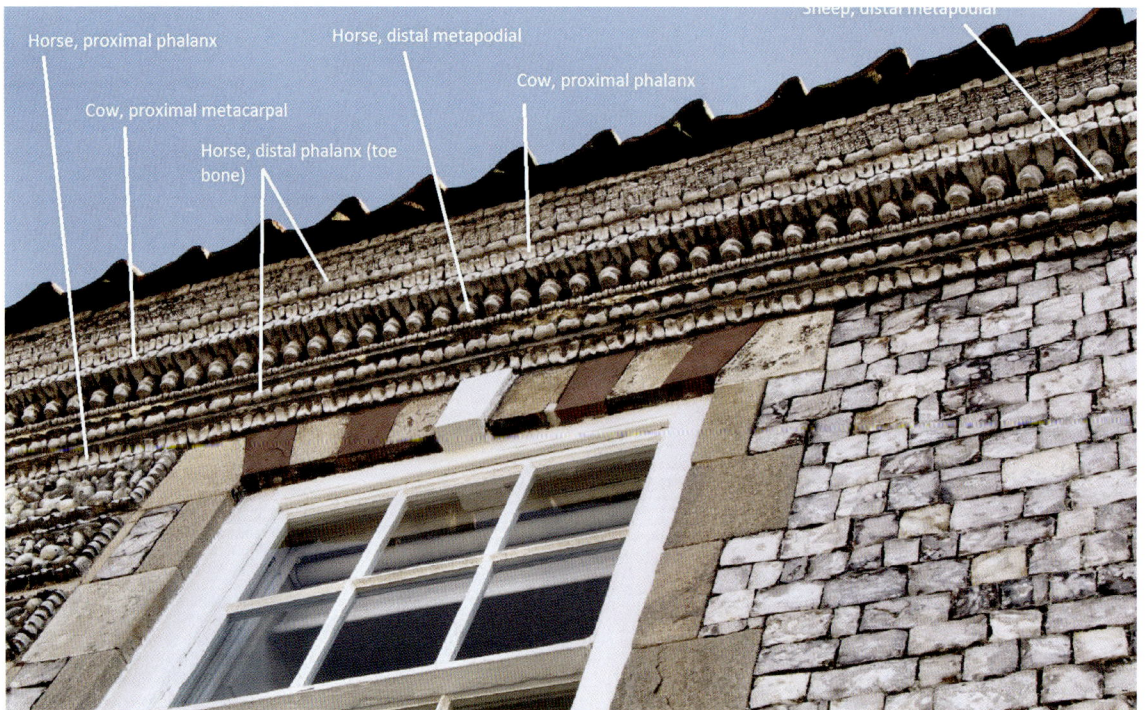

Horse, proximal phalanx

Cow, proximal metacarpal

Horse, distal metapodial

Horse, distal phalanx (toe bone)

Cow, proximal phalanx

Sheep, distal metapodial

Bone identification, Cley-on-Sea, Norfolk.

A flint and bone detail, Cley-on-Sea, Norfolk.

Cobble, gauged flush work and horse-teeth floor, West Dean House, West Dean, West Sussex.

Flint and horse-teeth floor detail.

Gauged flush and horse-teeth floor, West Dean House, West Dean, West Sussex.

Wooden block teeth repairs.

4. **West Dean Garden Folly** (West Sussex) – Although West Dean House itself was remodelled in 1804, some of the garden and garden features were completed at a later date. In line with the design of many gardens and features of the time (late nineteenth century), designers and patrons were heavily influenced by Italy and classical architecture; the owner of West Dean was no different. However, what does make this place unusual is the use of horse teeth and gauged flush work as a decorative finish on some of its folly floors. This may sound rather morbid, but both make sense as they are hard wearing under foot and would have been in abundance during the period. It is obviously quite difficult to know what or who instigated the idea. Using teeth or bone in buildings in that era was not unique. West Dean House itself is a quite remarkable flint building in design and technique. The window quoin detailing and galleting is quite exceptional. It was clearly built by highly skilled people. Perhaps it was one of these individuals that saw the potential in each material and put it into practice. Whoever it was, the level of work and use of contrasting material is very effective. The design was clearly also very well received, as it has been used in three locations in the gardens. Of note in the last West Dean image are the later attempts at repairs to the floor. With the absence of horse teeth, shaped and grooved pieces of wood have been used instead.

Chapter Five

Construction Planning

Construction Wall Types

The two most common types of flint wall construction are double-skinned freestanding walls and single-skin cladding, sometimes known as a 'faced' wall. Double-skinned walls feature a visible 'fair face' on both sides of the wall. The width and build up of these walls can vary. A typical example of this would be a garden boundary wall. Single-skin cladding is when a flint is built and tied onto a backing substrate. This can be applied on just one or both sides of the substrate. Depending on the project requirements, the substrate can vary from block work, reinforced concrete, timber or various other materials. Single-skin cladding is often chosen when there are structural or building regulations requirements. The substrate can be quantified and the flint is mainly decorative. Examples of this may include a retaining wall or a house wall.

Construction Methods

There are three main types of flint-laying construction techniques: freehand; shuttering; and pre-cast blocks and panels.

Freehand

Freehand is flint laid without the use of any temporary supporting aids. It is the purest form of laying, and provides the best opportunity to achieve optimum results.

The history of the method
It is very hard to say when freehand work first started. Perhaps it is reasonable to say in the time of the Romans. Their most common method of building would be *Opus incertum*. The process for this would be to build the outside of the wall first out of flint, brick or a composite of both, and then in-fill with a lime concrete and undressed chalk and rubble inner core. This is probably the closest method to current freehand laying techniques.

The process
The flints are laid one by one balanced on a horizontal bed of mortar, with a vertical mortar joint between each flint. The size of both the horizontal joint and the vertical joint will depend on the desired style or effect. The course above will then be laid on the subsequent completion of each course. Any vertical or horizontal line can either be checked by eye, string line, level or straight edge. Laying freehand can be critical if the wall or structure is curved, or if there is any detail such as flint size gradation or colour change. Freehand requires the most skill and experience of any flint-laying method.

Advantages and disadvantages
The freehand method offers maximum control and the ability to see the face of the flint that is being laid, together with the spacing of the adjacent flints. With the correct selection of flint and mortar consistency, height should not be an issue. It is the method most recommended and the process is explained in detail

further on in this book. I see no disadvantages to laying freehand.

Shuttering

Shuttering is a form of construction method where temporary boards are placed against the front flint face to support the wall whilst the mortar cures.

The history of the method

The technique is not a modern invention. There is evidence that the Normans first started using shuttering as a building method. The historical period of construction, together with purpose, were huge influences – a case of speed and functionality over aesthetics. It was certainly a period of time when numerous large fortifications and religious structures were being built over a short period; the majority of these would have been built by unskilled labourers, most likely overseen by a more knowledgeable eye. Walls of this period would often have been built up to 1500mm thick; add that factor to an often inconsistent method of lime burning (possibly resulting in a slow or poor set), and you can understand why the process of shuttering was adopted. One must also remember that this was also a period when speed was of importance. Thick, wide and tall structures were forms of protection. The Normans recognised the importance of building strong defences and fortifications. Defence shore forts were their main foothold in Britain. They built them in numbers, and they built them quickly.

There is other evidence to show that the functionality of a wall has nearly always been more important than aesthetics. The agriculture environment is an example of this. Many field boundary walls and agricultural buildings use this method. Their purpose was to manage livestock or store feed or grain. In these cases it is understandable why shuttering might have been the preferred option. These structures were often in remote locations, serving a functional purpose. Livestock do not care what a wall looks like. Material available might have also been influential, for example if small flint were the only available option. The time of year may have also influenced the type of construction: a slow set in the autumn or early spring of what would have been an air-trained mortar. Building field walls would be part of the farming calendar, as was the picking of the flints themselves. The main opportunity to build (when there were not other agricultural tasks to complete) would have been late spring and late autumn, avoiding the warmer summer months. These were good periods for a slow and gradual mortar set, but there was more chance of inclement weather.

There are some examples of shuttered walls being well built in an agricultural environment. East Tisted is a good example. Whoever built this internal wall managed to do it while retaining a very level and vertical face. The wall has been built using relatively small flints with a lime slurry pour. There is little evidence of the original shuttering support. It would have been almost impossible to build any other way with the materials used.

Shuttering as a technique continues to be used to the present day. The practice of shuttering is still regularly carried out, as it is often the preferred method of laying flint for bricklayers. This is not just down to ability and confidence, but also due to bricklayers being more comfortable in using a wetter and finer particle sized mortar for brickwork. Therefore the risk of slumping is much greater. It must also be said that some adopting the shuttering technique have become quite proficient at achieving a sufficient finish. However, It is a false and unnecessary belief that shuttering will provide maximum opportunity to gain height within a short period of time.

The process

The shuttering process involves placing and fixing a horizontal board where the proposed wall line is to be. Often scaffold boards or ply boards are used, but any timber with a straight edge is appropriate. Once in position, a horizontal bed of mortar is placed against the back face of the shuttering. One by one, the front face of the flint will be placed and pressed

The Rotherfield Park Estate

One quite understated example is the farm building walls on the Rotherfield Park Estate in East Tisted, a chalk downland environment in east Wiltshire. The majority of the local buildings are constructed of field flint with the occasional snapped quarry flint. Agriculture is the main industry and there are numerous agricultural buildings throughout the village. These may appear to be quite unassuming farm buildings; however, I believe that the flint type and size has been used both successfully and to its full potential. In most situations, the flints used would be thought of as unusable and discarded due to their small size and lack of face. However, I believe that whoever constructed the wall used the size and irregular shape to their advantage. They have created a dense interlocking structure that is standing up to the test of time very well; the wall has been constructed by shuttering and using a flint size of 20–75mm across the board. The flints are bonded together using a lime slurry. In my opinion the success of the build is in the execution of retaining such a flat and plumb face. There is no question that in terms of status this application would only be applied to a low-status functional building and not one designed to demonstrate success and affluence. However, from a flint layer's perspective, this shows more success in terms of ability and understanding of the potential of the material.

Shuttered field flint and lime slurry wall, East Tisted.

hard against the shuttering. A vertical mortar joint will be placed between each flint. Once the course is complete, the next course will be laid and so forth. When the top of the shuttering has been reached, another piece of shuttering will be placed in position and the process repeated. Normally depending on weather or the binder used, the shuttering will carefully be removed to expose the face of the wall. It is then that any unwanted or surplus mortar on the face of the flint or in the joint will be brushed back and removed. This final finish is not always the case. Often mortar is left on the front of the flint, or evidence remains of mortar that has oozed through the shuttering boards clearly showing the stages of the build.

Advantages and disadvantages

This practice normally eliminates the risk of slumping, but at the expense of the general aesthetics of the wall. It is often a process adopted with coursed work or any style with a larger mortar joint. There is not the need to see each individual joint or flint face that is required with finer work, such as knapped or random flush work. It would also be an impossible method if galleting is involved.

Pre-cast blocks and panels

These are made in a factory off-site by pouring normally cementitious mortar into a pre-defined mould. Flints are then pushed into the wet mortar to achieve the desired effect. The blocks or panels would then be transported to site and laid in position.

History of the method

Pre-cast panels were first introduced and manufactured on a grand scale in Britain after World War II. Due to the destruction during the war, large-scale post-war rebuilding and redevelopments were necessary to rehouse and rebuild the city's damaged infrastructures. There was a new building boom and, to cope with the current trends in design, the demand of this greater scale of manufacture required and the shortage of skilled craftsmen, pre-cast panels were the perfect answer. This was a quick, simple and cheap form of construction. Prefabricated panels and blocks could be manufactured and mass produced off-site, brought together on completion and erected at relatively high speed. It was a fast process that addressed demand and the development of both construction methods and building material technology.

Excess mortar on a flint face after the shuttering process.

However, it was not until the 1960s that the demand for speed, together with the fashion for function and form, addressed the renewed awareness and respect for vernacular materials and styles. Flints of all shapes and sizes were being used in pre-cast panels. The new levels of structural design stress that were now being introduced into buildings – particularly of a 'high rise' nature – influenced the first set of national building standards, formally known as 'Building Regulations', in 1965. These were a set of prescriptive standards and mandatory measures that had to be followed by all builders. These measures set out the minimum structural standards and performance of a number of materials, helping to enforce national rather than regional standards. Flint in its natural state created issues for structural engineers, with the new set of regulations and performance targets that they had to abide by. The issue was the individuality of each flint, making it almost impossible to calculate and predict the compressive strength of each flint and therefore the load-bearing capacity of a structure. However, this could be calculated if the structural part of a wall was either reinforced concrete or block with a decorative flint finish.

The process

Flint blocks also came about as a result of the post-war building boom. The manufacture of these blocks is similar to the bigger pre-cast panels. Made in a factory off-site by pouring normally cementitious mortar into a pre-defined mould, flints are then pushed into the wet mortar to achieve the desired effect. The blocks are then delivered to site and laid in a staggered bond similar to any type of normal block work or brick-work. Once the flint blocks have been laid, the whole area would normally (but not always) be pointed up.

Advantages and disadvantages

The additional repointing at the end of the process is designed to give a more pleasing appearance and hide

Prefabricated flint panels.

Flint blocks laid in situ.

the vertical and horizontal joint lines of the blocks. Unfortunately, what tends to result is the appearance of a thicker, well-defined and much too uniform joint line between the flints. Often it is very all too easy to see the exact dimensions and laying pattern of the original block. This issue has been recognised by the manufacturer and tradesmen alike and attempted to be resolved in two main ways.

1. **Adding smaller flints** – Some manufacturers and suppliers of the more conventional size blocks (440mm × 225mm) addressed the issue by placing smaller flints between the laid blocks to bridge the joint line. At one stage the manufacturer would even supply a bag of flints especially for this task. Despite good intentions, this rarely works due to the limited depth and space, resulting in poor adhesion and limited opportunity to add flints.

2. **Repeated notches** – Other manufacturers supply flint blocks with repeated notches to gain depth for the additional flint. The current trend would appear to be the manufacture and use of flint blocks of irregular shapes and different dimensions. These include both staggered and quadrilateral as well as curved and wavy interlocking systems.

Although all these methods and systems clearly attempt to address the issue of hiding the block joint line, they do not address the fact that the flints are laid in a factory off site with limited spatial awareness and consideration of the relationship between the flints within the block dimensions and the relationship between each block. This understanding, together with the selection of the primary face, forms the basis for good flint work.

Unfortunately, with a considerable recent increase in the number of flint block manufacturers, there clearly is a demand for this product. However, it is hard to say if this is just a result of market forces and a response to the current fashion of using flint in contemporary builds, as opposed to pre-cast blocks actually being a good product. A more worrying product on the market is imitation flint blocks. These have the dimensions of regular dense concrete blocks, but instead of using natural flint to face the block, moulded reconstituted stone with coloured pigment or dye is used. Once again, these imitation blocks can be laid by any bricklayer or competent tradesman. They tend to be repetitive in pattern and dull in texture. Though despite being no comparison to the real material, they are used remarkably frequently.

Why are flint blocks used? It is certainly not a matter of cost. Current research shows that the comparable cost savings using flint blocks over freehand work is relatively small. The manufacture and transportation of blocks also have a bigger environmental impact. However, it must be recognised that they certainly address some of the issues that are associated with flint work, particularly on the larger-scale projects or developments. There is a definite issue of skill shortages, and a requirement for speed on speculative projects. It also means that on large projects or sites, the main contractor can remain in control. There is more choice of bricklayers available to lay the blocks, making it not just more competitive, but easier for scheduling. As the blocks are manufactured elsewhere, the onsite element also becomes reduced. To a certain extent they also address any issue of weather conditions. Cementitious-base blocks also mean that they can be laid in most weather conditions (with perhaps the exception of snow and a hard frost).

However, the principal reason that flint blocks are used on domestic and commercial modern builds is to address the requirements of building regulations. Most modern building regulations require a cavity. This is normally to reduce the ingress of water and aid 'U values' retention (a measure of heat loss in a building element, such as a wall, roof or floor). Flint cannot be laid with a suspended face. This is a leaf of material that is self-supporting whilst retaining a cavity behind. Therefore, the need for a backing substrate is vital. With a separate backing substrate, it is crucial that the two material leafs are somehow adhered together. This is normally completed by the use of wall ties or mechanical fixings. It has been attempted, but with these in place it makes it almost impossible to lay the outer flint skin against a temporary internal shutter. The shutter cannot be removed due to the position of the wall ties. Whatever the dimensions of the substrate, any instalment will increase the overall width of the wall, creating possible cost and design implications. Therefore, a self-supporting flint block may address these issues at the expense of aesthetics.

Conservation restrictions and planning conditions have influenced the extent to which blocks have and can be used. However, this does appear to be rather erratic and inconsistent between planning departments. Some

Dyed reconstituted stone replica flint.

councils permit the use of blocks and some do not. There have been legal cases to both prohibit and give permission for the use of blocks. This book has not been written to debate or influence planning policy; however, it is important to address the issue of building a token flint block panel at the front of a new development to appease any planning conditions. With planning being so subjective and site specific, there is little or no opportunity for blanket rules. However it would be useful to have some form of consistency or quality control of the aesthetics of blocks, rather than just requiring them to meet structural standards.

Whether real flints or dyed reconstituted stone, flint blocks and panels have the nickname given by onsite freehand flint layers of 'the devil's work'. This is due to the poor aesthetic comparison that they have to traditional flint-work laying techniques.

Gabion baskets

Gabion baskets or cages are rectangular or square wire mesh boxes that are filled with flint and have been popular over the last twenty years. Historically used as retaining walls for erosion control, they became popular in domestic garden design after being used by an award-winning garden designer in 2000. I am not a fan of gabion baskets, as they rarely show the true potential of the material or the skill of the craftsperson; however, I do admire the ingenuity of the design.

The process

Once erected the gabion basket is filled with flint. Hand-placing of flint is recommended to reduce distorting the baskets. Depending on the project requirements and the budget, it is popular to have the flint just at the front of the cages and use an alternative substrate to either support the basket or to save money.

Method Considerations

Laying flint by the freehand method provides the best opportunity for achieving the optimum results. It should be the preferred technique of construction.

Flint in gabion cages in domestic build.

Flint in gabion cages used as sea defences.

However, a lack of confidence, ability or skill shortages are some of the main reasons people may opt for shuttering or flint blocks. Of particular concern for a flint layer is slumping.

Slumping

Slumping is when the mass weight of the material pushes the lower sections of the wall out. This bulge in the wall can not only push the wall out of the desired line, but also create weakness in the structure by leaving a cavity in the wall and reducing the bond between the outer leaf of flint and either inner leaf or backing substrate. Slumping is always avoidable. Most common causes are: inappropriate mortar consistency (especially too wet), laying too quickly for the previously-laid mortar to cure/set, and the flint not interlocking well enough with adjacent flints or backing flint.

Causes

There are numerous causes or increased reasons why slumping might occur. These include:

- Poor training and experience – this can lead to not selecting the correct face and bed of the flint.

- Poor planning and working – especially in too limited an area within one day.
- Poor preparation of the laying surface or substrate – insufficient preparation reduces adhesion and binding.
- The weather and atmospheric conditions – working when too wet or cold without protection of the wall will add weight and increase curing time.
- The thickness of the wall – this can result in increased weight, downward pressure and longer water evaporation and curing times.
- The style of flint work – for example, flush work will mean reduced depth and opportunity for bonding and adhesion to the side and rear face of the flint.
- The types of flint used – different flint types provide a range of opportunities for bonding and adhesion. For example, gravel pit flint or sea flint (cobbles) tend to have a smooth textural cortex, reducing the opportunity for any bonding or interlocking. On the other hand, field flint tends to be more irregular in shape and its cortex slightly more textured,

providing limited but increased opportunity for bonding.

- Type of backing substrate – any substrate with reduced suction will increase the risk of slumping. Plastic-based substrates in particular provide very little suction. Often on domestic new builds, cavity tray details or wrapping around steel posts will mean reduced bonding, therefore requiring further planning.
- Missing or too few wall ties – if laying flint against a fixed composite man-made substrate, insufficient wall ties will increase the chances of the two material leafs being held together.

Solutions

Though a lot also comes down to ability, patience, knowledge and confidence, there are many factors that can reduce the chances of slumping. These include:

- Appropriate preparation with a clean base or substrate – if building or rebuilding onto an existing structure, preparation should include the removal and consolidation of any loose material, making sure it is free from dust or loose mortar.
- The condition of the flint to be used – how much it has already been handled or worked, the supplying source, storage conditions and how long it has been stored will often define if the flint needs cleaning before use. Unworked chalk-quarried flint can often be delivered with an excess chalky powder of the cortex that is not an integral part of the flint. However, it is field flint that is often most likely in need of cleaning before use. Depending on the supply source and how long it has been stored, field flint will often be delivered coated in a thin layer of dried mud. This is relatively easy to remove either with a bucket of water and a brush or hose, or alternatively a pressure washer. Spreading out the flint on a hard-standing surface over a period of time can also work wonders, letting the elements do the cleaning for you. Over a

much longer period of time bleaching of the cortex will also occur.

- Good planning and setting out – giving a thought to planning the day's work. Where possible, spreading out the working area to increase length but reduce height. However, be aware that there is a fine balance between bringing numerous areas up but still retaining control in terms of finishing and protection. This can be an issue not just during inclement weather and winter works, but also during periods of accelerated curing due to direct sunlight or wind.
- Experience and practice in the selection of how the flint is laid – careful selection of the overall depth of flint and base platform is crucial. This is particularly the case at the start of the day and around the greatest load-bearing section of the wall. It is crucial that this section has a well-bonded flint and the mortar is well adhered to the rear substrate. If it is necessary to use flints with reduced depth or others with a suspect base platform, they should be used with caution and only at the top of the newly laid section at the end of the day. These flints are known as 'sliders' and should be used only when necessary and knowing the risk involved.
- The mortar, aggregate type, binder and consistency – it is not always possible to choose or select the type of mortar used as an existing match or if a precedent has already been set. However, it is important to be aware that different aggregates will have various particle shapes and sizes. The sea-dredged sands tend to be larger and more angular in shape and therefore able to hold and interlock better. Some of the finer soft sands tend to be rounder and smaller in size, increasing risk of movement.
- The consistency of the mortar in terms of moisture – the manufacturer's recommended water content should always be used. This can sometimes feel too little or that the mortar is

Flint colour tonal transition from black to white.

Colour-graded snapped quarry Flint for The Flint house project.

have fewer flints in their fields, as flint can damage machinery and reduce crop yield. However, long gone are the days when a farmer or landowner will let you have the flint for free. Most have become savvy to the increased demand, and therefore value, of the material. Many farmers have a stockpile of flint from fields they have already cleared and often will let you pick the flint yourself. However, it is essential that you always gain permission to remove flint from their field or land.

Flint picking

Flint picking from fields is best done at certain times of year when the flint is the most accessible. In some regions, flint picking was once part of the seasonal calendar; this activity was undertaken by agricultural workers, when there were fewer demands for harvesting or lambing tasks. Flint picking is best done in winter when there are no crops growing or sown in the field, just after ploughing and before rolling (compacting the soil over sowing). The ploughing action will often turn buried flints up to the surface, which makes the picking easier. Flint is after all the one crop that never fails. Flints can be picked from the surface, or semi-submerged flint can be levered from the soil using a small trowel or pointed hammer. The act of

flint picking closely echoes the mannerisms of the Renaissance painter Brueghel's peasants, engaging with their rural surroundings.

Picking flint by hand is also a good way of obtaining the exact size and colour of flint that you require. In Sussex, picking can be used to source coursed field-flint walls. Measuring is simple, with the flint size often simply gauged by a clenched fist. Any poor quality or unwanted flint can be left in the field, perhaps for another picker. It is important to make sure that you are picking in the right field for the material you need. Quantity, size and colour can all vary in different fields, or even different positions in one field. With ploughing movement and seasonal conditions, over time flint will spread across a field. However, as it originates in horizontal strata bands, the material is often denser in certain areas. With experience, you can become a proficient picker and know which fields and where in the field to source your desired flint type.

Once they have been picked, flints can be stored in bags. If you have more space available then you can also store the material loose on hard-standing ground. Storing the flint loose is preferable, as wet or windy weather can clean any remaining soil residue on the stones. However, if they are stored on soil, the flints can get dirtier from the splash-back off the

Stored field flint on the South Downs.

ground. As any dirt or soil will wash through the pile of flints with wet weather, it is worth turning the flints over every now and then to expose the bottom flints for cleaning. If dirty flints are still an issue, they can be washed individually in a bucket, or sprayed with a pressure washer. However, no cleaning will be required if the flints are free of soil when picked.

An essential part of surveying the landscape whilst flint picking involves field walking. Field walking can be a brilliant time to find the odd fossil, piece of broken farm machinery, or geological curiosity. One particularly common find is a 'shepherd's knee cap' or 'thunderstone'. According to folklore, the sound of loud thundercracks led people to believe that something more than lightning had buried itself in the ground. As a result, unusual objects found in the earth were often called 'thunderbolts' or 'thunderstones'. These included fossil belemnites, sea urchins and, more rarely, ammonites, pointed quartz crystals, lumps of iron pyrites, and Neolithic stone axes.

Flint Picking on the South Downs

I have many fond memories of flint picking, especially when the weather is fair. Searching for flints on the South Downs with skylarks singing above and eating lunch while leaning against a fence post is a calming experience. As you slowly blend into the local landscape, you become acquainted with the wildlife that surrounds you. During picking breaks, I have had some memorable wildlife encounters with stoats, foxes and even daytime badgers.

As it was also believed that lightning would never strike the same place twice, these objects became amulets for protecting the home. In early twentieth-century Sussex, fossil sea urchins were set on the outside of kitchen window sills and dairies to stop milk going sour, as thunder was thought to 'turn' the liquid. Similarly, ammonites were used in Wiltshire and Gloucestershire. It was reported that the mischievous village boys never interfered with them, simply because they were thunderbolts.

Chalk-quarried flint

Similar to field flint, the best options for obtaining quarried flint depends on the project and the quantity of material required. There are a number of suppliers online that sell chalk and gravel pit types of quarried flint. Depending on your requirements, both types can be bought as a raw material or as a worked product. Chalk-quarried flint in particular can come in different colours, translucency and rind thickness.

Quarried flint was often a by-product of local agricultural lime suppliers or cement works. Due to decreased demand for agricultural lime, smaller quarries were amalgamated into larger ones, and many closed down as their quarrying licence or natural source ended. Many smaller quarries also had a change of use and became landfill sites. You may find local sources for small quantities of chalk-quarried flint. Due to the increase in flint value, some chalk quarries have prioritised flint over agricultural lime supplies. Others have heavily invested in facilities and machineries to match the demand for flint.

Hand versus machine

It is worth asking if the quarry flint is worked by hand or by machine, as some of the larger suppliers will only sell flint split using pneumatic cutters. As the demand for worked quarried flint has increased, so has the use of these machines. It should be recognised that there has not always been the skill base available to supply this new demand with hand-worked flint. However, these machines are particularly relentless. The flint is fed into the machine and when the operator is happy with the alignment, a blade is released to split the flint. This may happen numerous times until the required shape or size is achieved. Although skill and experience is required to operate the machine and predict how the flint may

break, there is still a lack of control, which limits the quality of the finished product. As a result, it is often obvious if a flint has been split by a machine or by hand. Interestingly enough, it is not necessarily at the expense of speed.

For chalk-quarried flint, you can buy direct from some of the larger quarry companies. As with the field flint, you can also purchase the material from suppliers that use the flint for other manufacturing processes and are prepared to sell from their stock supplies. The same weight and quantity restrictions apply. Unlike field flint, quarried flint is easy to buy from builder's merchants, although normally at a premium. Around 80 per cent of builder's merchants buy their quarried flint from a single supplier in East Anglia. This quarry supplier stocks flint from two of its own quarries, which is great for material consistency but harmful for the local landscape.

As mentioned previously, in addition to different types of flint there are also regional variations of the material. Consequently, the majority of recent flint new builds across the country are built with East Anglian flint. This variation is often known as 'milk flint', due to its grey, non-translucent appearance. Though there is nothing wrong with this kind of flint, it often clashes with the vernacular. As a result, the repetitive use of one flint type can make the character of buildings too uniform and obscure the local distinctiveness of flint. Hopefully, the increase in demand will mean greater choices of flint source and variation.

Gravel pit-quarried flint

Gravel pit-quarried flint is very different to chalk quarry flint in regards to options and availability. For both the projects and the suppliers, the majority of gravel pit flint work is found in East Anglia. The suppliers are numerous, although they often change as quarries open and close due to merging or un-renewed licences. The colours, translucency and quality of gravel pit quarried flint can vary. The quarries are open for the digging and supply of sand

and aggregates. The actual usable flint is a by-product of the sand and aggregate process, and usable gravel pit flints are known, priced and supplied as rejects. Some quarries will have very little usable rejects, but others have a plentiful supply. The colour of material can vary from quarry to quarry. Some are known for their white flint stone, but some known for the more sought-after black flint.

Makers of imitation prehistoric artefacts will often go between quarries looking for individual pieces of flint. Therefore, the location of reliable quarries that supply high-quality workable flint is often kept relatively secret. Unfortunately, this is the same for architectural flint knappers. The open sourcing of skills is mostly encouraged within the architectural flint industry and there is recognition that there will be enough work for everyone. However, sharing information about sources of the raw material appears to be a different thing. This is understandable though, as when a desirable quarry is identified the demand can skyrocket. For example, recently the location of a particular high-quality black flint became public knowledge and almost immediately the prices rose and the quarry ran out of its quota of the flint.

Although preserving the regional vernacular by using local flint should be encouraged, often this is not always possible. There are currently just a few options countrywide of locations that supply high-quality quarried flint that is usable for finer status work like flush work and quoins. If it is chosen well, gravel pit flint can be a good alternative. Most of the time, an intensely worked flint will have its rind removed, which reduces any indication of its origin. This means that if there is no local quarry available, it is possible to match the inner core of certain flints to others from different regions. However, this can only be done with a well-worked flint where the finished product contains no indication of either rind colour or flint. If you look closely, the iron deposits

Stored chalk-quarry flint, Hertfordshire.

in a gravel pit flint can be the only subtle telltale sign that it is not local material.

Sea cobble flint

As with all other types of flint, sea cobbles, beach cobbles or sea flint can come in different colours, shades and translucency. Under the Coastal Protection Act 1949, it is illegal to remove stones from public beaches. In fact, some areas of the country feature large signs warning people that they could face prosecution if they remove stones. Therefore, buying sea cobble flints from legitimate online suppliers or via a builders merchant are the only options. Packaging and delivery are the same as for chalk-quarried and gravel pit-quarried flint: 1-ton bags delivered by pallet or bulk deliveries of 10 tons minimum. Depending on the density of the flint and joint size, a 1-ton bag will cover approximately 7 square metres; however, this will vary on the size and quality of the cobble.

Types and sizes

In East Anglia it is possible to match the vernacular (local stone) with material from gravel pit quarries. Both the rind colour and inner colour will vary depending on the quarry, but the majority will be shades of brown and yellow. Brown-, yellow- and pastel-shaded rinds are often popular hues for projects. Although the flint cobble colour is not an issue, it can be more challenging to obtain bulk quantities of set sizes. This might explain why most cobble work in East Anglia is in a random style. This is similar to the 'penny bun flint' found by the coast of Dorset and East Devon.

The cobble work on the south coast is often more coursed and gauged. Unlike gravel pit flint, it is possible to buy graded cobbles. They are dredged, beach-picked and graded in locations on both sides of the English Channel. French tend to be blue or grey in colour and English with more variation, including browns. Long gone are the huge quantities of south coast cobbles that would be transported to the potteries for the ceramic industry; these were once sourced and picked by hand. Now only harvested by machine, in the UK they are still graded by hand. There are commercial suppliers located in positions to select optimum-size cobbles. These locations are chosen by beach location and longshore drift direction. The optimum locations on both sides of the English Channel have not changed, although most cobbles are now imported from France.

Sea cobble flints are normally sold in 75–100mm sizes, 100–150mm sizes, 120–200mm sizes and random sizes. The colour will vary from browns to greys, with the latter the historically preferred hue. Colour will depend on the source. In England cobbles are still hand graded; however, in France this practice has stopped within the last ten years. Do not use decorative stones as replacements. The 'Scottish decorative pebbles' from the garden centre that are sometimes used are often the wrong colour and are inappropriate for this function. If you are completing coursed work, the consistency of cobble size is crucial, as the technique can be unforgiving.

Quoins and specials

Quoins, flush work or flints used for any decorative work are normally only available by special order from either flint knappers or conservation building suppliers. Most of this material will be

Stored gravel-pit flint, Norfolk.

Graded and stored
cobbles.

worked by hand using hammers and then copper punches for the final finishing. This shaping can also be achieved by machine operators, though these are still often finished by hand. The machined quoins can be problematic, as they are rarely good enough quality to achieve a perfect continuous angle. They are also less tactile, which is one of the main attractions of the worked flint. This may be the result of mass production (due to a requirement for speed and quantity).

Quoins and flush work can come in different colours, shades and translucency. They will most likely be made from either chalk-quarried or gravel pit-quarried flint. Occasionally they can be made from field flint, but as most field flints are unpredictable to work, it is harder to produce consistently finished items. Quoins are normally completely free and clean of rind. This is despite the fact that some quoins, particularly when abutting a field flint style, look better with a proportion of rind remaining. With a small semblance of rind, they 'read' better as an entity and there is less aesthetic separation between the quoins and adjacent flint body.

Types and sizes

Quoins can either be random or gauged. You will need the latter if you are abutting gauged work. These are typically in a 70 or 75mm gauge and are sold either per linear metre or per unit. The gauged knapped work typically comes in at a height of 75mm. If used in conjunction with stonework or brickwork, the height can be altered to the appropriate height and joint size. It is rare but not impossible to buy flush work in a bigger gauge, at least in large quantities. It is also possible to make and therefore buy square flush work, though normally it would come in varying lengths to achieve optimum coverage per worked flint.

Historical picking of cobbles, Normandy coast, France.

A tray of flint gravel-pit quoins.

Different lengths will also aid a staggered bond when laying (not having any stacked vertical) and is often sold by the square metre. It is important to specify how worked the flint is, as gauged work and joint size can vary considerably. If possible, it is best to request a sample; however, this might need to be paid for.

Gallets

Gallets are usually only available by special order from either flint knappers themselves or conservation building suppliers. They are ordered by volume in either buckets or bags. Depending on the density of the galleting, 10 litres in volume will complete 3 square metres in an average joint size. Similar to the quoins and flush work, gallets are often made from chalk-quarried or gravel pit-quarried flint. Occasionally gallets can be made from field flint, but the unpredictability of working field flint means that it is difficult to get thin and usable gallets. Gallets can come in many colours, sizes and styles. As they are made to order, it is essential that you know which size, style and colour of gallet you require. There is a widespread belief that gallets are just waste material. This is true to an extent, but there is a lot of waste that is unusable for galleting, as they must be of a certain size and depth. If the material fragment is too blunt or long, it becomes an issue to push into the mortar or lay the gallets close together. From experience, many galleting projects require you to make the gallets, especially for the contracted work. This can be undertaken off site when all the materials for the contract are being prepared.

Reclaiming and salvaging

Your selection of flint will inevitably vary from project to project; however, if you are repairing or rebuilding an old wall, then reclaiming the flint is recommended. This is normally a simple task if the flint has been laid in a lime mortar. Doing so will not only save you money and the hassle of finding a suitable match, but most importantly it reduces waste and environmental impact. 'Reading the flint' (understanding the best face to use) is clearly defined by the clean face of the flint and the faces discoloured by contact with the old mortar, though this can rely on the skill level of the original flint layer. Depending on the type of wall and quality of mortar, normally up to 80–90 per cent of the original flint can be reclaimed.

The process

When salvaging flint, it is usually best to start by breaking the wall down into smaller pieces. This task can be completed with a small hammer-action breaker or hammer and bolster. Once in smaller pieces, any excess lime mortar should fall off the flints with the strike of a hammer. However, be careful not to hit the flint too hard, as this may lead to it shattering or breaking. Just a light hammer blow can cause enough shock vibration to separate the mortar from the flint. Do always remember to wear the correct PPE when performing this task, as both flint shards and mortar particles can fly off dangerously on hammer impact.

Reclaiming flint from a cement-based wall is a completely different procedure. Despite good building intentions, it can often be virtually impossible to salvage enough flint to make it worthwhile. The average flint reclamation rate for a cement-based wall can be as low as 20 per cent of the original flint. As cement-based mortar is so hard, it not only takes longer to break a wall into smaller pieces but much more flint gets damaged in the process. Even when the flint is broken down and individually cleaned, there always seems to be some unsightly cement mortar residue left on the flint.

When reclaiming, you must be aware that your structure may have different styles of flint on either side of the wall. A north- or south-facing wall may have the same flint finish, but have a different weathered patina. If you do choose to salvage the flint, you must be organised by storing any different variations in material in separate bags or piles. Matching old flint to reclaimed flint is also possible, but depends on the type and quantities of flint that you require. With an old wall the original source is most likely to be close by, whereas with field flint it would be maybe worth approaching the local landowner or farmer. With chalk- or gravel pit-quarried flint, obtaining a perfect match from a local source may prove more challenging, as the number of local chalk quarries has diminished hugely.

Selection of mortar

The design of flint-work mortar (for example, selections of aggregate and binder) will have a major influence on a mortar's performance and appearance. First, you must consider what the properties required from the mortar are. To select the correct materials it is important to contemplate the requirements of the flint style and structure, the environmental conditions, the aesthetics and the materials involved. You must identify whether you are designing a new mortar or matching a pre-existing mortar. If you are matching a historical mortar, the sand, binder and any additives should be re-created in your binder type, particle size

Tubs of prepared gallets ready for use.

variation, colour and proportions. This is, of course, unless the pre-existing mortar has clearly failed! Some historical mortars have traditional additives such as coal particles, ash lumps, brick dust, earth, animal hair and chalk particles. It is important to ascertain the role or function of these inclusions. They may have been included for numerous reasons – some intentional, some unintentional – due to the lime-making process. If you are designing a new mortar, you must consider the build's function, vapour permeability and aesthetics when deciding whether you should match a historical precedent for the type of flint work and location.

Mortar design should not come down to material availability or lack of knowledge and experience. Whenever possible, I would always recommend designing the mortar and sourcing the constituents yourself. This would be through a process of selecting the right materials for the application in hand. This may require time researching and understanding the existing or proposed structure and selecting materials that have the suitable required characteristics. Some lime mortars are commonly available in pre-mixed form ready to use or just require water to be added. Although these may be convenient and reduce variation issues, they are often too generic in composition suitability. For production purposes, many pre-mixed mortars have a limited size and characteristic range. They may also contain additives. Some specialist suppliers may be able to provide bespoke mortar matches, which may prove inordinately expensive.

Historical lime mortar in a coastal location.

Aggregate

Aggregate is a general term for sand, gravel and other materials. These particles can vary in size and shape. They can give the mortar both body and increased adhesion. They can be natural, manufactured or recycled. Sand is a natural granular aggregate that derives from the erosion of rock and surface material. It is defined by the size at the lower end of the particle scale (4mm and below), whereas gravel and grit are defined by the upper end of the particle scale (4mm and above).

A good sand for flint work is one that is sharp (angular in shape) and clean from salt, clay or contaminants. A high clay content will impact mortar performance by reducing binder adhesion and encourage cracking and shrinkage. It should be well graded and contain an even balance of all particle sizes. A poorly graded sand is when there is a limited variation in particle size. Ideally, the predominant particle shape should be irregular in form. This is because it is the interlocking of the irregular shaped particles that give the mortar structure and strength. This shape also enables better opportunity for binder adhesion, and in turn produces a stronger bond. The source of the sand can be from a sea, inland pit or a river. Depending on its origin, the sand may be available in a huge range of colours, particle shapes and sizes. It is important to remember that sand is a natural resource and therefore may be subject to some variation. This can be both regionally and within a single pit. Although most pit sands can vary in colour depending on the source, there are some staple sands that remain fairly consistent.

A range of sands and aggregates.

Aggregate sourcing

The choice of aggregates plays an important part in flint laying. This is an element of the building process that is often not given enough time and consideration. A well-graded chosen aggregate will have a major impact on not just the colour and visual appearance, but also the physical property and performance of a wall. An inappropriate use of sand will be problematic during laying, structurally inferior and aesthetically unpleasing. There is a diverse range of aggregates and sands on the market; however, unfortunately all too often the flint worker just accepts whatever is supplied and delivered by the local builder's merchant.

The use of different sands and aggregates reflects the local availability of various materials. Historically, this would often be as a result of transport limitations. There is evidence that a lot of historical sands had less consistency of grading, were not washed and had a higher clay content. They tended to be finer, with a more isolated inclusion of larger particle sizes. Due to the modern transport system and economic pressures, sands are now frequently brought in from long distances. Pit- or sea-sourced sand is currently the suppliers' preferred choice of source. Most modern sands have a reduced but more consistent particle size. There is a lot more colour dying, enhanced crushing and blending of the material to achieve the desired particle proportions and colour. Due to the British Standards regulation of sands and aggregates, it might be necessary to blend sands in order to create an appropriate match for an historical sand type.

Sand and aggregate supply has always been down to geography, availability and economics. Modern supply is now much more about availability (quotas and pit leases) and economics (transportation and profits). With an improved transportation system, aggregate and varying regional demand, aggregate is no longer locally sourced and supplied. As many suppliers are changing from small local pits to large pits, variation is reduced. The grading demands of modern building methods, particularly those of the concrete and asphalting industry, have also influenced the type and variation of sands and aggregates available. To keep up with this demand, and due to limited pit quotas and leases, supply is going more towards the vast supply of marine resources.

It may be worth considering that there are a number of other reasons why the sand at your local builder's merchants may be that particular sand. Seasonal demands of the construction industry also influence

A clean, well-graded, sea-dredged sharp sand.

An example of poor aggregate selection.

aggregate supply sources, movement and variations. In the case of marine-dredged aggregates, it may be surprising to know that it can take less than 72 hours between sandbank and bag. This is less to do with the freshness of the sand, but more to do with running an efficient dredging fleet. The seasons and time of year can also influence where sand can be taken from and therefore the type. Marine sandbanks are all mapped out and worked within a pattern. This is to reduce damage and impact on marine life. Certain sandbanks cannot be dredged during the shellfish season. Therefore, as very little marine sand is stockpiled, there can be huge variation from one month to another. This is all worth considering if you require consistency of materials for a large project.

The names of sands will also vary from region to region. A 50:50 pit sand from Hertfordshire, a coarse sharp sand from Sussex or a coarse sharp washed river sand from Devon can be a similar colour and have the same particle size, depending on origin. The same sand in one region may be called different names by different merchants or suppliers. In Sussex, sharp sand, compo, coarse sharp and coarse concreting sand are all the same product.

Aggregate selection

To increase the chances of a successful project outcome, it is important to select the correct sand or aggregate. This can be subject to the binder being used, the task involved, historical precedents and the build location. Depending on the source, particles can be either round or sharp. Although there is evidence that many historical sands were finer with little shrinkage, this should not be mixed up with particle shape. In the case of modern mortars, for sake of workability and convenience there is a tendency to use a finer round particle size sand. These mortars with rounded, same-size sand particles rely more on the inherent strength of the binder to hold them together. It is poor practice and a misconception that increasing binder content will improve adhesion and strength. Unfortunately, this choice can result in an aesthetic mismatch with historical sands, creating a reduced-strength mortar and causing issues when completing the task.

There is a clear correlation between particle size and joint size. The maximum sand particle size in lime-based mortars should be about one-third of the width of the joint size. Historically, mortar joints were tighter. A good balance of particle sizes will control the

Dredged marine-washed sand being graded at Shoreham port.

even spread of binder through the mortar joint. When a sand of poor particle variation is selected without any adjustments or blending with other sands, there is a temptation to add additional binder or water. However, this is poor practice and can produce a reduced-strength mortar or increase the risk of shrinkage. Sand adjustments without blending can be successfully achieved by hand sieving or large mesh screening.

Although there is often evidence of the current and historical use of aggregate type, there is presently debate about the risks of using salt-laden sea sands. These sands potentially react with certain binders, which lead to slower setting times and increased risk of holding moisture. A fair amount of my own work takes place close to the sea, and therefore we regularly use sea-dredged sand. However, the source is carefully selected to be free from salt and contamination. This is locally known as coarse sharp sand, marine-washed sand or compo. It is a well-graded sand type that has a cross section of particles from fine up to 4mm for medium washed sand and up to 6mm for coarse washed sand. Not only is it confusing that there are different names for the same product, but the colour and particle size can vary depending on where the sand has been dredged. Some sandbanks along the south coast can vary in colour from pale to yellow, and have varying particle sizes. As mentioned previously, an ideal sand for flint work has a good cross-section of all particle sizes. Therefore, in order to match the inconsistent grading and inclusion of larger sand particles in historical sands, we will often add a small proportion of large aggregate.

Lime

Lime is made from burning limestone, a naturally formed sedimentary rock whose principal constituent is calcium carbonate, at high temperatures. The kiln process drives off carbon dioxide. It is also possible to produce lime using other naturally occurring materials that are high in calcium carbonate, for example seashells and marble. Lime is used as a binder to mix loose sands and aggregates together and produce a lime mortar. The lime reacts with carbon dioxide

in the air to eventually form a hardened state, similar to what it originated from. The aggregate can provide structure, body and strength. In addition to flint wall mortar, lime is used in other building processes (such as plasters, renders and paints) and structural masses (such as floors and earth structures).

Benefits of lime

The production of lime and its use is well established and the material was probably brought to Britain by the Romans. The numerous benefits of using lime as a binder make it ideal for a number of functions in construction. There is a huge range of appropriate and affordable lime products currently available on the market, which are suitable for all different applications. Whether it is being used in new builds or historic building work, lime is one of the most versatile and workable of materials in the wet building trades. Due to its flexibility and versatility in application, lime mortar is the preferred material to use for all types of flint work. Following good practice lime is not difficult to handle. It is a shame people often lack confidence in the product or are sometimes scared and confused about the different lime products on the market. This is because, if it is correctly selected, applied and cared for, lime can offer long-term benefits that cannot be matched by modern cementitious substitutes.

Using lime is regarded as being less damaging to the environment than using other binders, such as cement. This is because the production of lime requires a lower temperature than cement, and therefore needs less energy to manufacture. Furthermore, as lime mortar cures it reabsorbs carbon dioxide. This process offsets the carbon dioxide released during manufacture and makes it a carbon-neutral product. In comparison, cement products only absorb up to 20 per cent of the carbon dioxide produced during the manufacturing process. Although cement does contain lime and limestone (calcium carbonate and calcium hydroxide), its non-crystalline nature traps free-moving carbonates, which reduces re-carbonation and therefore becomes an 'end of use' product. However, though lime does have the benefit of being close to a carbon-neutral product, progress still needs to be made. To achieve net zero carbon emissions, improvements need to be made to manufacture more efficient equipment. Most kilns used to burn lime are gas- or coke-fuelled kilns that were developed for the cement industry. Therefore there is some debate that, although lime is produced at lower temperatures than cement, a poor process and limited equipment efficiency actually has any real environmental impact. Also huge improvements could be made on: the use of natural hydraulic limes (NHLs); finding a source of a suitable national alternative (the majority of hydraulic lime is currently imported from the continent); reducing the impact of transporting sands long distances (for economic reasons); using more local suitable sand. The reduction in the use of pre-mixed bags of lime mortar would reduce the unnecessary further production processes, transportation of raw materials and the transportation of the finished product.

It is only recently that we have become increasingly aware of the damage that some modern materials can inflict on buildings and structures. This is particularly the case when workers are seduced by the speed and convenience of cement-based products. It is now widely acknowledged that these impermeable products are completely inappropriate for historic building work. Any structural movement takes place in the mortar joint and *not* in the structure's main component. Therefore, using a flexible lime-based mortar will allow for limited movement. As lime mortars are more breathable, they can reduce and transfer trapped moisture and condensation. Lime mortar is known to be 'self healing'. In certain limes, water can transfer and move uncarbonated free lime particles to places where it can carbonate (that is,

Various types of lime in different forms.

exposed cracks). This is known as 'free lime'. The amount of 'free lime' will depend on the quality of the lime itself, together with moisture movement within the wall.

Once you are aware of lime's availability, flexibility and sustainability, it is hard to understand why someone would choose any other product for their mortar mix! Just because lime has been widely painted as a dangerous and corrosive material, this does not mean you must use alternatives. Lime is similar to so many other products or practices: over-exposure or misuse can be harmful, but following good practice will reduce most of the risks. Therefore, it is important that lime-based materials are well understood and used with confidence.

Lime selection

Forty or fifty years ago there was a lot of talk and excitement about a 'lime revival'; however, lime usage is clearly here to stay. I was no lime expert when I first started using lime, and I am still no lime expert now. However, all those within the wet building trades must be open to adapting, learning about, and sharing our practice. This goes for getting to grips with lime, too.

Although building limes tend to be associated only with conservation work, in my own practice I use lime for all flint work, including both new builds and historic structures. The only difference is the type of lime used. There are now so many types and brand names of building lime products on the market, it is unsurprising that not just novices but also experienced designers and practitioners of wet building trades can find the prospect of finding the right lime daunting. Although NHL limes, especially the default NHL 3.5, are very out of favour – especially in comparison to hot lime – I believe there is a place for both, depending on the circumstances. This is normally split between historic building work and new builds. The important part is to understand the physical properties and performance characteristics of each lime and then select the suitable match for each application

– and not choose just for convenience or to fall into default mode.

As well as there being different types of building limes, using different aggregates and additives will create mortars with a range of characteristics suitable for a variety of applications. It is not a case of one lime product fits all tasks and situations. It is important to understand the differences in performance and application. Although industry experience has shown that some limes appear inferior as a product, all building limes can become inferior due to misuse or application.

When choosing a building lime, you must understand the task in hand. You need to be able to identify whether you need to use a lime mortar where no precedent has been set, or if you need to emulate an existing historical lime mortar. In terms of plasticity and permeability, the mortar that you use must work in the same way as any existing mortar. Unless the historical mortar has failed, it should be 'like for like'. If the mortar you use is different to the original, it may not only be aesthetically unpleasing but can create future issues for the existing building or structure. If selecting a new mortar without any precedent set, it is important to consider the water absorption and vapour permeability required by the structure.

It is also crucial to understand the materials to be used. The mortar should always be softer than the materials being laid. Do not forget that flint is a much harder material than bricks and most stones. Therefore, with the assumption that you are not matching an existing historical mortar and the need for maximum permeability and flexibility is not a priority, this could mean the use of hydraulic limes is more feasible. Also consider what the structure is, or what it is constructed of. If it is a new build, is it a masonry or timber construction? If you are building against a timber substrate, a lime with more flexibility may be more appropriate.

Take time also to reflect on the environmental conditions of the task. Air limes need access to carbon dioxide in order to set. Therefore, if you are working in a wet environment (such as water bridges, sea defences or canals), unless you are using a pozzolan to increase the hydraulic set, maybe hydraulic limes are a better choice, as they can set underwater. In more hostile environments (such as exposed locations, copings or chimneys), the decision might rely on both the type of lime but also the strength of the lime.

The time of year at which you are completing your project might also influence the lime selection. Using appropriate practice and protection, it is definitely possible to use air limes throughout the year. However, their slower setting times and vulnerability to frost make them more problematic than hydraulic limes. It is also possible to use pozzolans to accelerate the set on air limes, but this may change the mortar characteristics enough to jeopardise any mortar matching. Again, it depends on the task being undertaken, the quantities of lime required and the location. If an air lime is the most appropriate match for a structure in a very exposed location, scheduling may come to the forefront. As a result of this, many flint practitioners schedule certain jobs at certain times of year.

Types of lime
Type 1: Natural hydraulic lime
What is it? Natural hydraulic lime – NHL – is produced by slaking quarried limestone containing reactive silica impurities, such as clay, and other minerals. It is supplied in powder form in sealed bags. Variations in the raw material and burning process can produce hydraulic limes of very different characteristics, including varying strengths and colours. Although its main set is by a chemical reaction when in contact with water, some setting is still attributable to the carbonation process. It is categorised by its compressive strength in laboratory conditions after 28 days. The three main strengths are NHL2, NHL3.5 and NHL5. They are sometimes associated with the terms 'feebly hydraulic' (NHL2), 'moderately hydraulic' (NHL3.5) and 'eminently hydraulic' (NHL5). These terms relate to the amount of active

clay materials they contain. The more hydraulic a lime is, the quicker it sets and the higher its final strength. However, this also means that it is less breathable and flexible. Although these strengths are European standardised, there can be huge variation at which end of the specific classified NHL scale they are on depending on the manufacturer. The NHL strength will also vary depending on the sand to lime ratio, the aggregate used and how well it has been mixed.

Availability? Hydraulic limes are available to purchase in sealed bag form ready to be mixed onsite with a chosen aggregate. They are also often available to purchase as pre-mixed bags. These formulas come blended with aggregates in the form of dry, ready-mixed mortars, in either small bags, ton bags or mortar silos. They have become very popular due to their convenience, availability and marketing. It must be noted that there are currently no makers of any NHL limes in the UK. Most hydraulic lime is imported from Europe, with France, Portugal and Germany being the main three countries providing the material.

Advantages? NHL lime sets quickly and predictably. As its chemical reaction sets with water, it quickly reaches an initial set reducing risk of frost damage. It is probably the most forgiving option for inexperienced users. It is great for new builds and good for use underwater.

Disadvantages? Hydraulic limes have reduced flexibility and moisture control compared to non-hydraulic limes. Strong NHLs come at the expense of permeability and flexibility. In addition, some of the stronger hydraulic limes are completely inappropriate for use on historic buildings and may cause more damage than good. There is too much variation in the NHL compressive strength spectrum between brands, which are a variation of free lime. Some brands have no free lime, when others have a high free lime content. An increased hydraulic set means extra care and protection is necessary to retain moisture and slow the process down in the summer months. There is a risk of encouraging a fast set for increased productivity at the expense of the mortar. In winter months, slow or failed carbonate due to water saturation and low temperatures will make them vulnerable to frost damage and produce inferior characteristics. If NHL bags are not sealed and stored in a dry place, any atmospheric moisture or direct contact with water will initiate the setting process before use. Hydraulic limes tend to be less workable to use. The practice of using gauged limes or hybrid limes, that of mixing lime putty to hydraulic lime to improve workability, produces a product of unknown characteristics. There is a higher environmental impact of lower carbon recapture and production and supply compared with air-trained limes. All NHL kilns in Europe are fuelled by coke.

Type 2: Non-hydraulic lime

What is it? Non-hydraulic limes are also known as air limes. They are given this term because their main set is by carbonation and drying when in contact with air. This chemical reaction enables the lime to set and harden. Air limes are produced by burning quarried pure limestone. This process removes carbon dioxide and produces a dry, lumpy (often referred to as lump lime) product called quick lime. Quick lime is a highly reactive material that should be handled with caution. Variations in the raw material and burning process can produce non-hydraulic limes of very different characteristics. Differences in the raw material and its source can also result in differently coloured non-hydraulic limes. When quick lime is added with water, the process is called slaking. The impurity content (active clay materials) of the quick lime changes the slaking time and the expansion during slaking. These are classified as 'fat limes' or 'lean/ thin limes'. Variations in the slaking process can produce different mortars of variable quality; hot mixed lime mortars and lime putty mixed mortars being an example of this.

Availability? The five main non-hydraulic lime products available are hydrated lime, pure lime putty, pre-mixed lime putty mortar (with added sand and aggregate), quick lime and hot mixed lime mortar. Depending on the process or quantity of water, slaked lime can be supplied in powder form in sealed bags and sold as hydrated lime or mixed with water and supplied in plastic tubs and airtight lids, and sold as lime putty.

Type 2A: Lime putty and lime putty mortar

What is it? Lime putty is a non-hydraulic lime mortar. It is also known as fat lime or slaked lime. If a specified amount of water is added and retained in the slaking process, it produces a creamy textured slurry material called lime putty.

Availability? Lime putty is sold in either pure putty form ready for mixing onsite, or pre-mixed form with aggregate already added. Both are normally stored in plastic tubs with a layer of water on top to prevent any carbonation or chemical reaction until they are used. Well-stored and maintained, lime putty and premixed lime putty mortar should have an indefinite shelf life. In fact, it is commonly accepted that the longer a putty is kept in storage, the better the 'quality'. This includes improved aggregate particle adhesion (acting as a binder), plasticity, workability and water retention. This is also known as mature lime putty and sold in pure form or pre-mixed. The pre-mixed formula is often known as 'coarse stuff'.

Advantages? Lime putty mortar is extremely soft, flexible and permeable. Until the recent resurgence of hot mixed limes, it is these characteristics that made it the 'go to' for use in matching historic lime mortars, plasters and paints. It creates a very good bond between binder and aggregate. It offers good workability, producing a mortar that adheres well to masonry. It is very permeable and cheaper than NHLs. Lime putty mortar offers the closest match to historic limes in terms of a bigger range of slaked particle burn. It is easy to add a pozzolan to increase the hydraulic set.

Disadvantages? As it relies on both the evaporation of water and the carbonation process, lime putty-based mortars can have a slow set time. Although this can often enhance the performance quality of a lime mortar, depending on the task undertaken or the skill of the tradesman, this can also be problematic. The set is dependent on many factors, including practitioner skill levels, the materials involved, substrate, thickness of mortar, temperature and atmospheric conditions. To overcome some of the atmospheric issues, it is common practice to add a pozzolan. This natural additive will encourage early strength and setting characteristics, reducing risk of frost damage and slumping in load-bearing tasks, although this is now less of an issue with improved manufacturing processes. A poorly manufactured quick lime may produce over-burnt or vitrified particles that increase the slaking time, maturing times or cause future issues of carbonation 'popping' in a mortar. This can result in exposed, isolated raw lime lumps that can produce uneven curing and be aesthetically displeasing. Although once common practice the production of lime putty by slaking dry hydrated lime will produce greatly inferior characteristics and performance to that of one made directly from quick lime.

Type 2B: Hot mixed lime mortar

What is it? The current most fashionable lime in use in historic building work! Its use has recently had a huge revival in the industry, though the process has been around for a long time. Recent research and analysis exploring the durability of Roman concrete has found that it was down to the Romans employing the hot lime process and not as it was previously assumed the choice of specific pozzolanic materials.

Hot mixed lime mortars are made by adding the desired aggregates and sands during the slaking process. The slaking process produces a hot chemical reaction, hence the name hot mixed lime mortar. There is a public misconception that this process has to happen each time before immediate use and that the mortar has to be hot whilst using.

In fact, it is the process of slaking with the sands and aggregates that is important, not the timing. Although only the wet trades have only recently re-engaged with the use of hot lime, there appears to be no difference in final performance whether it is used hot or cold.

Availability? Hot lime mixes can be made on-site using bags of quick lime and chosen aggregate. They can also be purchased as a cold and pre-mixed (aggregates added) wet mortar and can be stored until needed.

Advantages? Hot lime mixes create a very good bond between binder and aggregate. Good workability, producing a mortar that adheres well to masonry. They have high free lime content, resulting in residual expansion in voids and joints (self healing); compared to lime putty, this improves the ratio of lime to sand, making it cheaper. They have a relatively rapid initial set and are very permeable, while being cheaper than NHLs. They are also the closest match to historic limes in terms of a bigger range of slaked particle burn. The heat generated may aid the incorporation of additives and their reaction. It is easy to add a pozzolan to increase the hydraulic set.

Disadvantages? All limes are caustic and can cause issues in incorrect use or if PPE protection is not used. Quick lime is, in particular, a highly reactive material. The intense exothermic reaction during the slaking process can be dangerous if it is not completed in a controlled manner. Although an initial fast set, due to the high lime content and slow curing process, hot limes can be prone to failure due to frost. However, the 'free lime' content in hot limes is deemed a positive characteristic, as it is relatively soluble, but it can be prone to unsightly leaching if subjected to a continuous flow of water. It can be removed before it carbonates, or even better still, uphold good practice by using a pozzolan to increase the hydraulic set or by protecting the mortar during construction and reducing saturation. If you are inexperienced, it is too easy to add too much water in the slaking process and 'drown' the mortar.

Type 2C: Hydrated lime

What is it? Hydrated lime (often known as 'bag lime' or 'builders lime') was once the most popular of limes due to its availability, over-use in design specifications, and a lack of alternative lime knowledge. It can produce slight differences in strengths, but its range is nothing compared to the spectrum of natural hydraulic limes. The use of hydrated lime with cement does not create a lime mortar, rather a cement mortar with increased workability and improved flexibility.

Availability? The product is available in most builder's merchants and sold as a powdered form in sealed bags.

Advantages? I see little advantage in using hydrated lime when there are a number of alternative superior limes readily available in the marketplace.

Disadvantages? Hydrated lime has declined in popularity for a number of valid reasons. In theory there should be no issues with freshly manufactured hydrated lime; however, it is unlikely to be available. Although it is sold in sealed bags, the fact that its chemical set is with air means that it is possible for the carbonation to start during the bagging process and continue throughout the storage period. Unlike non-hydraulic lime in putty form, the longer it is stored before use, the more inferior it becomes. With the introduction of superior lime alternatives and more user-friendly plasticiser alternatives for the general building trade, less hydrated lime is sold and used. In turn, extending the period the bags may be on the shelf of a supplier further increases the risk of carbonation. This is often known in the trade as 'dead lime' or 'blown lime'. When a 'dead lime' bag has been opened, the lime is no longer in powder form but in one solid lump or lots of small lumps. Continuing to use the lime in either of these inferior forms can lead to a flawed lime mortar. The mortar is flawed because the binder will not be creating an even bond between all the aggregate particles to hold the structure. Also, over time, carbonated lumps can be exposed by the elements that produce unsightly leaching of white streaks; this is sometimes known as 'wall bleeding'. Hydrated limes tend to be less workable to use. The practice of using gauged limes or hybrid limes – that of mixing lime putty to hydrated lime to improve workability – produces an unreliable product of unknown characteristics.

Pozzolans

Pozzolans are materials that can be added to air lime mortars to speed up the initial setting process and increase compressive strength. Their presence will react with non-hydraulic lime to help initiate and enhance the hydraulic set. As a regular feature of historical mortars, pozzolans are still in common use today. The main difference between historic and modern pozzolans is that in the past they were added as part of the aggregate content and were therefore more visible. Modern pozzolans tend to be finer and can now be used as part-replacement for the volume of lime; they also tend to be the same colour, making them less apparent. Replacement proportions will vary depending on pozzolan, lime and performance requirements. They can be both artificial and natural, and come in different forms. Some are clearly identifiable in historic mortars, some less so. They are usually readily available from most conservation building suppliers. Historically, popular pozzolans include volcanic ash, tile or brick dust and wood and coal ash. Modern artificial pozzolans include ground granulated blast furnace slag (GGBS), HTI powder (a product of high temperature insulation) and metakaolin (produced from calcined china clay) and pulverised fuel ash. Recent discussion has looked into the benefits of using NHLs as a pozzolan. The choice of pozzolan will depend on the aesthetics and hydraulicity required for a project. It is important to note that some pozzolans are more reactive than others. The hydraulicity of all pozzolans can range depending on variations in raw material firing temperatures and particle size.

The selection of pozzolan or cautiously increased pozzolan proportions will aid the hydraulic set underwater or in a wet environment. This can also be useful if the task at hand requires an accelerated set to avoid frost. However, this outcome is not guaranteed unless good practice is followed.

Cement

From the early twentieth century, Portland cement almost all but replaced lime in construction. Portland cement is produced in a similar manner to lime. It is limestone that is heated up, except heated to much higher temperatures and with clay additives. In the early nineteenth century it was not possible to achieve and retain such high temperatures; therefore, unlike today's cements, they tended to be relatively weaker and more permeable. Modern processes and higher firing temperatures now produce a much more unforgiving rapid-setting, harder and impermeable product, the use of which is unsuitable and inappropriate in many situations.

There was a period when, due to the slow setting time of air limes, it was common practice for hydrated lime to be gauged with high cement content. Unfortunately due to convenience, lack of knowledge and understanding, this specification would be the norm. Often to meet the desired aesthetics and create the visual illusion that mortars were only lime based, there was a period of time when the use of white cement became popular with air limes. Although this may create a more aesthetically pleasing finish, it would be at the expense of less compressive strength and issues caused by accelerated setting times. There are now clear signs that its use has caused accelerated masonry deterioration in old buildings and use with historic mortars. Though cement increases the speed of the initial set, it comes at the expense of flexibility and permeability. This makes it completely inappropriate for use on lime-built historical buildings or structures. A high cement content in lime mortar

An example of an historical pozzolan used as part of the aggregate content.

Examples of historical and modern pozzolans.

can reduce mortar performance by weakening it through 'segregation'. During this process, the binder's aggregate adhesion is reduced and it is not equally distributed throughout the mortar.

Any cement content in the mortar will also reduce the 'working window' when laying flint. This 'window' is the opportunity to use the mortar to lay flint or undertake and complete any finishes to the mortar joint. Due to the hardness of cement-based mortars, it is almost impossible to reclaim materials at a later date. If rebuilding a wall, you may expect to salvage a high percentage of flints bedded in a lime-based mortar, but only very few if they are bedded in a cement-based mortar.

Pigments and mortar

There are many elements that can affect the colour of the mortar, the main two being the choice of aggregate and the choice of lime. The use of a pozzolan can also be very influential in colour variation. Although modern pozzolans tend to be neutral in colour, historical pozzolans such as ash or brick dust can create strong colour tones. Whenever possible, it is important to try to match like-for-like in colour and texture of sand, lime and any use of pozzolan.

There is a current trend, especially in new builds, to use a dark mortar with random snapped flint. This is not new – it has been a popular combination since the mid-eighteenth century. It made the joint sizes smaller and raised the status of the build. The main difference between the two building periods is the current use of chemical dyes and pigments, compared to the historical use of industrial by-products and natural materials. The danger with using modern chemical dyes is their

Phoenix Sand, an Example of Historical Mortar Quirks

We were once having to complete some major repairs to a thirteenth-century church in East Sussex, which had been through a number of structural design changes. We were completing remedial tasks, particularly addressing the failure of some of the mortar to a nineteenth-century addition. It is a common practice to undertake a mortar analysis and produce mortar samples before the work commences. Through a visual analysis of this particular church's mortar, high-density black particles were identified. Although I consider myself well versed in the use of pozzolan and pigment additives, these were different. The existing mortar appeared pale in colour and there was little need of black particle volume for a pozzolan. Curious but confused, I sought the advice of a local practitioner who had given up the trowel some time ago. With great amusement, he explained that it was called 'phoenix sand'. The sand was a waste

Phoenix sand.

product from the (now closed down) local iron-works, where it was regularly used as moulds in the foundry process. There was a deal that 'spent' sand was available cheaply or free to the local wet trades: as a result, phoenix sand was used only in a specific area surrounding the old ironworks.

building, analysis may be required. Depending on the project you are completing, this examination can be undertaken in a variety of ways.

Analysis companies
It is possible to send off a mortar sample for analysis. If this is the route that you choose, you must make sure that you select the correct mortar sample to send off. Has the existing structure been repointed? Has the wall been historically altered in any way? Taking three samples would be recommended to guarantee consistency. One of these samples should include the inner core of the wall. However, it is important to note that some historical flint mortars may have been part of a two-stage process. This is evident in the 'snail's creep' style that has different sand particle sizes between wall body and pointing effect, and also in some nineteenth-century dark mortars where pigment was only used on the visible outer layer.

Beyond simply identifying the mortar contents, some analysis companies may also supply an aggregate match or a pre-mixed mortar match. However, there are often disparities between the mortar identified in a sample and what is available locally. If it is at all possible, always try to send your mortar analysis to a local service, as they will have better knowledge of local sands. This may also give you the option of mixing the aggregate and binder on site, saving money and minimising excess waste.

Independent visual analysis
Matching the aggregate
If you are a confident and experienced flint worker or a novice, sometimes a visual analysis may suffice. First, you should make visual observations of the existing mortar contents. Sometimes the historical weathering of the mortar joint may remove some of

Poor mortar matching and selection.

Poor lime and aggregate ratio.

the binder, making mortar analysis easier. Make a note of the sand or gravel colour, particle size and shape. You might also note the proportions and distribution of the particle sizes (if they are different). Are there any notable additions? Shells, crushed stone, fuel or manufactured items? Is the mortar translucent or lustre? Is it dense or friable?

Next, you should crush some additional removed samples and complete the same visual analysis process again. Similar to when you select samples for analysis, be selective with where your mortar samples are taken from. It is important that they accurately represent the original mortar that you want to match.

Once you have performed your second visual analysis and made notes, you have the option to go one step further by oven-drying the sample, soaking the crushed material in diluted hydrochloric acid. You can weigh the sample before you begin the process. After the mortar sample has been soaked in the solution for a period of time you can wash it with clean water, being careful to retain all particles (including those fine ones suspended on the surface). Finally, the last step is to oven-dry and re-weigh the sample. This process should remove the mortar binder and make it easier to identify the sand and gravel particles. If all the binder is not removed the first time, it might be necessary to repeat this process. A stacked sieve system may

also give a clearer indication of particle size, colour and ratio percentages. If the before and after weight has been measured accurately, it may give you an indication of the aggregate to binder ratio. If you are inexperienced, be cautious during the analysis process as the acid process can dissolve any possible calcium (shell-based) particles that were part of the aggregate makeup rather than the binder. This means the cleaning process can remove a proportion of the fine content. This problem can be resolved by using particle filters in the cleaning process.

Please note that the diluted hydrochloric acid solution can be dangerous if used incorrectly. Often sold as patio cleaner or brick cleaner, it is a very corrosive chemical that can be harmful to you and the environment. If you do need to use it, take care and follow strict PPE guidelines. It is also advisable to use the solution sparingly and dispose of any diluted excess liquid responsibly.

Once you have identified the aggregate, you must then find an aggregate match. To get a good match, you may need to be prepared to visit a number of aggregate suppliers in person. From experience, suppliers are often not responsive to a phone enquiry requesting an aggregate blend of specific proportion percentages. Be aware that you will have to invest some time in a mortar-matching task. Doing so will benefit your project outcomes: ensuring structural integrity and the historical accuracy of the build's aesthetics.

Sand and aggregate can be obtained from builder's merchants, aggregate merchants and conservation building suppliers. Conservation building suppliers will often consistently stock aggregates from the same source and this limits the particle colour, size and shape. On the other hand, general builder's merchants are often supplied and stocked on an economical basis and therefore may vary. Do not make the assumption that all branches of builder's merchants stock the same aggregates or that they will consistently be supplied by the same source. In addition, do not assume that the aggregates stored in 25kg bags are a match to those sold loose. They can often be supplied by a different source.

Do be prepared to blend aggregates to achieve your desired matching balance. If you are successful in finding a matching aggregate, it is worth bulk ordering the material for your project. This is because if you do not have enough product to complete the project, you may have issues with aggregate changes and variations when re-purchasing.

Matching the binder

Once you are selecting the correct lime for the task in hand, you will need to determine the correct binder to aggregate proportions. It is sometimes possible to follow the manufacturer's recommendations, although proportions will vary from the requirements of the task, the choice of aggregate, any use of pozzolan and, if required, to match an historical precedent. Although sometimes this can often be worked out through experience, it is sometimes necessary to work out the binder to aggregate proportions for a specific task. A good way to measure the correct lime content required is to check the void content of the sand. This is called the void ratio. Void ratio will only work if the binder is in powder form. Depending on the sand shape, size and variation, the gaps and voids within the particles can vary. Too little binder in these gaps and voids will create a weak mortar. An overfill of too much binder or moisture in these gaps and voids will reduce particle adhesion and therefore encourage shrinkage and cracking.

A simple field test to determine the void ratio is as follows:

- Dry a small quantity of the sand in question.
- In two measuring containers, place equal amounts of sand in one and water in the other.
- Next, pour the sand into the container with the water. Agitate the container to enable the water to permeate the sand. Then let it settle.
- The void ratio is calculated by volume of water before the added sand, less the volume of water above the sand, once added.

A range of mortar biscuit samples.

A range of sample panels made to release planning conditions.

Once you feel happy with the choice of lime and selection of sand and aggregate, the next stage is to prepare some biscuit samples. These are small mortar samples that can be used to either match an existing mortar or view the aesthetics of a designed mortar. It is an opportunity to try different proportions to ascertain the most suitable match. These can be made in small squares, or use some cut-down plastic down-pipe to retain the sample. It is important to be systematic and clearly mark each sample with the contents and ratios. For each sample it is also important to complete all the finishing processes to achieve a true final finish.

With the chosen flint style and mortar it can be advantageous to build a sample panel to check all combinations of material. This is often a requirement of planning or listed building consent conditions. A sample can either be completed on a small scale for fine work or larger if required. I would also recommend building the larger sample panels in a movable box rather than in a fixed location. From experience there is never a right place when static.

Flint and chalk sample panel in progress.

Chapter Six

Tools to Use

The Principal Tools For Flint Laying

Mixing the mortar

There are various tool options when it comes to mixing mortar. The choice will often come down to the mortar used (type of lime, aggregate particle size and pigment), availability and quantities required. It is always worth investing in good-quality equipment that will get the job done, but depending on your project the use and investment may vary considerably.

For this reason a single, specific type of mixing equipment will not be recommended in this book, but advantages and disadvantages of each type will be considered. Some equipment does mix mortar better than others; however, it is important to be practical and realistic for what is achievable in different scale projects. Mixing equipment options include: pan mixers; cement mixers (normally 90 litre capacity); site mixers; plasterers; swivel stick mixers; and mixing baths. Whichever piece of equipment you use, it is good practice to keep it clean. Lime may provide a longer window for cleaning before curing, but it can be just as stubborn as cementitious materials when it comes to removal.

1. Pan mixers, roller mixers or forced action mixers – These probably have the optimum mechanism to guarantee a well-mixed mortar, particularly when using hot lime or lime putty. However, they are quite specialist pieces of equipment. Hire availability for these types of mixers can be an issue and they can be expensive to buy. This option would certainly not be suitable for all situations and people. As the mechanism is flat, some people may find it difficult to unload mixed mortar and clean the equipment.

2. Regular size cement mixers – There are various brands of these mixers on the market. They come in a variety of sizes, with the most popular size having a drum capacity of 136 litres (but a mixing capacity of 90 litres). There are also different power source options: 110v, 240v and petrol. Which power source you use would ultimately be decided by the location the mixer is most often used in. These mixers tend to have just two internal blades so mixing time will increase compared to the forced action mixers (which often have four blades). Different mixer brands and models can also have different stands. Stand choice may be dictated by the type of mortar that will be most frequently mixed. Lime mixes can be very sticky and most flint-laying mortars are considerably drier than a brick or render mortar. This can mean that if you are not careful, the mortar can stick or sit at the back of the drum (*see* the section on making mortar). A 90-litre capacity mixer is often a popular choice because of its wheel type. The drum can be tilted at various angles depending on the mortar mix or at what stage of the mixing process you are at. For example, the tilt may vary when you are dry mixing, adding water or mixing with all the correct quantities.

3. Site mixers – These are mostly for large-scale use on sites when a large quantity of mortar is required due to the size of the wall or number of workers. Most site mixers have a gated

blade design, which improves the mixing process and increases mortar quality through aeration. Site mixers also feature a multi-position lockable drum helping to reduce build up or mixing at the back of the drum. Due to the 250-litre drum capacity size, both large (110 litres) and well-mixed mortar batches of NHL mortars can be produced. Often, workers are too optimistic about how much mortar can be mixed in any size of drum mixer. However, drum volume is not the same as mixing capacity. Over-filling any size of drum mixer will lead to a poorly mixed mortar and create future issues. Site mixers are good at effectively increasing mix capacity, though they are not for the faint hearted. Like any piece of equipment, if used incorrectly they can cause serious injuries. Consequently, some workers choose to use two smaller drum mixers simultaneously to increase mortar capacity. Both methods work, the main emphasis being that good practice is essential: you should not reduce the mixing time or overload the mixing drums if increased production is required.

4. Paddle whisks – This mixer is also known as a plasterer's whisk. It can be useful when only very small quantities of mortar are required. Paddle whisks can be used as part of a large task-specific piece of equipment, or connected to a drill for smaller-scale mixing. Both involve mixing the required materials and quantities in a bucket. Paddle whisks are not only great for mixing small quantities, but if used correctly guarantee a well-mixed finished product. This is good if you are using either hot limes or lime putty in small quantities. Some putties or finer powders in particular can be hard to mix in a drum mixer. Therefore, if large quantities of these are required then a flat roller mixer or paddle mixer may be your best option. Although paddle whisks are flexible and reduce the amount of equipment needed to carry around, they can result

in over mixing and poor batching. The more batches you need to mix, the more likely you are to mix incorrect ratios and therefore have an inconsistent finish.

5. Mixing baths and boards – Both of these mixing options can be bought specifically to suit the task at hand. Mixing boards are often produced in an octagonal shape and have a lip to reduce mess and mortar spillage. Similar to the paddle whisks, these can be useful for small-scale mortar production. Hand mixing has been used for centuries; however, it can often lead to inconsistent mortar. The length of time that the mortar has been mixed and how well it has been combined is crucial to a usable outcome. Like bucket mixing and paddle whisks, mixing baths can result in unmixed lime or aggregate stuck in the bottom or corners of the mixing vessel.

If you only need a mixer for a one-off project or very occasional use, hiring may be a better option than buying. From experience, most hire shops have good availability of mixers and offer reasonable hire charges. If you are planning to use your mixer for multiple or long-term projects, buying new or buying on the second-hand market can be a good option. However, if you do buy second-hand, you must check the drum is clean and the paddles (inside the drum) are straight with a good connection with the drum. This is because the weld between the paddle and the drum can become weak during the cleaning process on a well-used or poorly maintained mixer.

In regards to mortar mixing methods, it is important to remember that one method does not necessarily suit all tasks or projects. It is vital to consider the practicalities, what mortar you are mixing and what quantities you need. Your working rhythm is also important. If you are completing a solo project by mixing mortar *and* laying flint, you must make enough mortar per batch to have a decent laying time. Even the most experienced layers take time to get into the flow of a project.

Profiles and block lines

The use of profiles, guides or string lines is important to obtain a consistent flat plane when using flint. As mentioned in the section on construction methods, this is one of the reasons that shuttering is chosen as the preferred process of choice. However, please note that though this method may create a flat face, it is often at the expense of aesthetics and the flint bond.

Laying freehand with a profile, guide or string line should always be the preferred method. The challenge lies in identifying which guide is appropriate for which flint style. For example, if the flint work is coursed, a horizontal string line would be recommended. If you are laying flints between brick or stone pillars, the use of corner line blocks or pins works well. If no end pillars are available, it is possible to construct wooden formas that will set the required profile of the wall. Once they are set in place, these will enable you to use and follow the string line at the correct course line and height.

Straight edges

Another favourite of many flint layers is the use of a straight edge. This is often used in conjunction with a string line. It can be particularly useful when laying field flint and can be used to either pull flints to the front edge or for levelling off the tops of a laid course. Most straight edges used are made of aluminium. There is no such thing as a 'flint layer's straight edge', so a plasterer's 2m-long aluminium straight edge is normally used. These are light and handy to use for many aspects of flint wall construction. Other uses for a straight edge include levelling out, striking (marking) lines and completing render capping.

Trowels

There is not one trowel that fits all, or one trowel suitable for all tasks. Most flint layers will have a number of trowels depending on the task or type of flint laying style. Yet, whatever the style or task,

everyone has their own favourite trowel for laying flint. These may vary from a 6-inch steel-forged gauging trowel (curved on the nose), a 6-inch pointing trowel and occasionally a 9-inch bricklaying trowel. I find the larger trowels or trowels with a point harder to control and work around a curved flint or random joint, although that is completely my personal preference. If possible, it is important to spend a bit more money on a good-quality trowel. This is because flint mortar tends to be much stiffer (due to aggregate size and water volume), which can place a large strain on the trowel and result in the breakage of cheaper brands. Other trowels that are often used by layers are bucket trowels and different sizes of tuck trowels. These can be very useful for pointing, but also for the variety of cutting required for pointing styles and finishes. The above suggestions are not exclusive. Open any flint worker's toolbox and you will find a number of tools that have been either adapted or made for a specific purpose. For example, a 15mm copper pipe can be cut and shaped especially for snail creep pointing.

Hammers

The hammer and the trowel are the two most important items for laying flint. It is therefore unsurprising that flint workers can be pretty obsessive about their hammers! After all, it is an essential piece of kit. Flint is a very hard material: it is ranked a seven on the Mohs scale of mineral hardness, as it is a form of quartz. The Mohs scale measures the ability of one natural sample of mineral to visibly scratch another mineral. When handling minerals, it is important to be aware of the hardness of your material in relation to the object you are using to work it. For example, if a hammer is too hard it can become brittle and susceptible to internal stress and ultimately shatter. Neolithic man worked out that without the option of using metal, antlers were the perfect balance for working flint. For modern hammers a similar balance is required. This is done by the correct tempering of

A range of flint-laying trowels.

A range of mixing and infill trowels.

the hammer for flint and is often why copper rods are used for more detailed flint working. Tempering, in metallurgy, is the process of improving the characteristics of a metal, especially steel, by heating it to a high temperature, though below the melting point, then cooling it, usually in air. The process has the effect of toughening by lessening brittleness and reducing internal stresses.

Hammers can vary, depending on the intended task and personal preferences of the layer. Even the difference between flint knapping and flint laying can require different hammers. The latter tends to need less of a range of weights and shapes. Useful hammer types can include:

1. Pick hammer – A pick hammer is very effective for the removal of unwanted mortar or foliage when preparing a wall.
2. Quartering hammer – Used by many gunflint makers, a larger quartering hammer (normally 3 or 4 pounds in weight) is required for the first hit on a flint. This hammer can be used to split the flint in order to assess its quality, or reduce it into a more workable size. Some knappers will opt for a plain claw hammer, others for a lump hammer. My personal preference is for a heavier stone mason's hammer (square at one end, beak-shaped at the other) with a longer shaft. This is because you can place your hand at the top or bottom of the shaft. These options can help you manipulate the size of the arc when swinging your hammer and create different knapping results.
3. Ball pein hammer – Although some flint workers like to us claw hammers for finer shaping of the flint, I know many that will use a range of sizes (8oz–16oz) of ball pein hammers to complete any final working or shaping of the flint whilst laying. These hammers are particularly useful for random snapped work or flush work.
4. Bricklayer's hammer with a scutch comb – These can be useful for cleaning flints or preparing a wall.
5. Copper rods can also be used. These are used for finer detailed work, when making flint quoins or artefacts. Copper rods are softer than steel hammers. They also tend to grip the flint and can be more precise for working on specific areas of a flint.

A range of flint-knapping hammers.

Copper rods used for finer knapping techniques.

Boards and buckets

Storing and working with mortar on site is an important part of the laying process. Bricklayers tend to use spot boards. These are boards that are raised off the ground to a suitable height to avoid repetitive crouching and bending down. Spot boards are usually spaced out at equal distances along the wall. This makes it easy to transfer the mortar from the board to the desired location on the wall, without blocking a clear access route.

On the other hand, many flint workers use and work their mortar from rubber buckets. These tend to be heavy-duty 25-litre buckets with side handles. The use of mortar buckets may be habitual, but has some practical benefits. In both brick- and flint-laying, you need to turn the mortar over to increase its workability and cut the correct amount of mortar for the desired joint. However, flint mortar tends to be drier

than brick mortar. As a result of this, the flint layer can spend too much time 'chasing the mortar around the board', whereas the wetter brick mortar can be turned over but stays in the same position. Keeping and working the drier mortar in a rubber bucket means that it is easier to press the mortar against the edge of the bucket to cut the required mortar quantity for the desired mortar joint. In addition, the mortar tends to be self-contained, which reduces the risk of soil, flint chippings or waste contamination. The bucket also makes it quicker and easier to move the mortar around the working area.

Rubber buckets also offer a practical solution to storing and moving flints when working besides a wall. Whereas bricks and blocks can be neatly stacked, flints are notoriously difficult to store because of their irregularity. Large quantities of flint can be transported and stored on site in 1-ton bulk bags; however, the required quantity of flint is often transferred to the buckets and refilled when necessary. This means when working along the wall (if right handed), you should have a bucket of mortar on the right side and a bucket of flints on the left side that you move along as you progress. If adopting this method, it is important both to invest in strong heavy-duty buckets and also make sure you are not overfilling your buckets.

Gauging buckets and boxes are also useful on-site tools. Both are used for measuring material quantities during the mixing process. Quantities are usually measured by volume rather than weight. If you are measuring and loading materials with a shovel, quantities can vary substantially (despite good intentions).

Brushes

Brushes are a surprisingly important part of the flint worker's toolbox. Different brushes enable flint layers to achieve different finishes. Consequently, flint layers are often very passionate about the advantages and disadvantages of wooden handles or the demise of quality brushes due to the rise in steel prices. Various brushes can be used for the different stages of flint work, from the preparation to the finishing.

1. Soft bristle brushes – These are good for preparing a surface and for cleaning up afterwards.
2. Hard, natural bristle brushes – This type of brush is sometimes known as a churn brush. They can be useful for preparing a wall as well as compacting and driving mortar into a joint. You can use a hard, natural bristle brush for the finishing 'rubbing up' process, which involves cleaning the laid flints and removing any aggregate particles in the mortar joint. Compressing a mortar joint with a hard, natural bristle brush means that the joint remains open. This tool allows the joint to breathe, as moisture can easily transfer through internally and externally. On the other hand, using a metal trowel or tool can lead to continual compression of the joint.

This can result in the 'fat' being drawn to the surface, thus closing the mortar joint. This may reduce water ingress, but can also trap water and make the joint more vulnerable to saturation and freeze thaw.

3. Wire brushes – The wire brush is the most used brush in finishing techniques. There are numerous types on the market, with the most popular being steel or brass. Wire brushes are available as two-row, three-row and four-row brushes. Most flint workers will have a collection of all sizes of steel brushes and perhaps a couple of brass brushes (these provide a softer finish and reduce the damage of softer materials such as sandstone). Steel brushes will not scratch or damage flint; however, they can remove excess mortar covering or surrounding the flint. Brush row size will depend on joint size.

Additionally, brushes are useful tools for obtaining the aesthetic effect of ageing. If used correctly they can remove the smaller aggregate particles and softer binder particles, replicating the weathering process. However, if this is performed incorrectly, they can mark and gouge a well-built wall causing irreparable damage. To create this ageing effect, a combination of tools and brushes work well. You can start with a larger four-row wire brush, completing the task with a two-row wire brush, and finishing off with a hard bristle brush to remove any evidence of brush marks.

Water spray and hessian

Another tool worth having on site is a water pump spray. This has a range of uses: helping with cleaning and preparing a surface, reducing accelerated mortar curing and enabling specific finishes. Using a light spray over a semi-cured mortar will remove the binder and extenuate the aggregate particles. However, this technique must be undertaken with caution. If the water spray is used prematurely, it can wash the mortar out and cause slumping, collapse, or mortar saturation.

Tool Anecdote

A favourite tool story of mine comes from my early days of laying flint. In the same area that I started flint work was a skilled old boy. He was a supportive yet discerning flint worker and we occasionally crossed paths professionally. Though he was always keen to see how my skill level was progressing, he was not afraid to remind me of how far I still had to go. At the time I was working on my first snail's creep project (raised pointing that meanders around the laid flint) and he was very curious to see how I handled it. Snail's creep is notoriously challenging and involves using fine-aggregate mortar that is pushed into the joint and then cut to create either a point or a convex shape. Any pointing tool you use tends to catch or drag any larger aggregate particles. I had completed all the flint laying and it was time to start on the snail's creep. My peer knew this (he had been keeping an eye on my progress) and appeared behind me inquisitive and grinning. He let me sweat and curse for a bit. I was conscious of being watched and struggling to achieve a neat and consistent cut with whichever trowel I used. After a while he got tired of watching and asked what I was having for lunch. I responded abruptly, telling him I was having a sandwich. He chuckled to himself and disappeared, returning with a fork, which he wordlessly offered. What use was a fork if I was eating a sandwich? With great joy he rolled the outer fork prongs, split the middle prongs into a 'V' shape and sharpened the fork on a nearby brick wall. He handed me my new tool. It was of course the perfect size and shape to complete the snail's creep pointing.

A roll of hessian can also be a helpful material to have in your toolbox. Its main purpose is for protecting completed flint work. Many people assume that protection is only needed for frostbite during winter months, but damp hessian can also reduce accelerated mortar curing in the heat of the summer.

Adapting Tools

The tools mentioned above are not an exclusive or exhaustive list. A flint-worker's tools often depend on the task at hand and personal preferences. It is important to also note that though many tools can be bought, adapted tools can often be valuable and occasionally more effective for achieving a specific technique or finish.

The tool anecdote above highlights the importance of experimenting with adapting tools. Your local tool shop is not going to have a flint tool section. Some tools have to be made or adapted to suit. For example, many years ago 'bucket trowels' did not really exist and most flint workers had to buy a 9/10-inch bricklayer's trowel and grind the end off to create a flat end that was more suitable for scraping the bottom of a bucket. If you cannot find a tool for a specific task, then you must get creative. Your flint work will benefit because of it.

Chapter Seven

Flint Knapping

Although there is a similar understanding of the material and approach required to work flint, there is a definite distinction between architectural flint knapping and knapping for experimental archaeology. The main difference will be the tools used. Most architectural knappers will use steel hammers, whereas knappers of experimental archaeology will use hammer stones, antlers and sometimes wood and bone. For finer work and pressure flaking, both will use copper rods of varying size.

The other main difference is that architectural knapping only involves direct percussion technique (blows aimed directly at the material), whereas experimental archaeology uses both direct percussion and indirect percussion (impact and pressure created by a punch and hammer).

Tools required for architectural knapping are relatively simple depending on the level of knapping required. Most important is correct PPE (personal protective equipment). Good goggles and gloves are a must. Tough footwear, preferably with steel toe caps, is preferable. Whenever possible, always knapp outside or in a well-ventilated area. If knapping for any length of time – even outside – it is worth considering a good-quality fine-particle face mask. Remember, due to silicosis, the average life expectancy for a gunflint maker in the early nineteenth century was only 44 years.

Also, choose your location well. Working flint will create a fair amount of waste. This is often sharp and can be difficult to clean up. It is also quite surprising how far an area of waste flint can spread, particularly during the early days of learning. I would recommend working on a smooth hard-standing surface so that waste can easily be swept and shovelled up. If working on grass, lay a board or sheet down first. In no particular order children, pets, lawn mowers, livestock and flint knapping waste do not go well together. Beware!

A range of hammers will help the different processes, from quartering and splitting to shaping. Weight, shape and shaft length can vary depending on individual preference. A hammer with a longer shaft will give you more options to help you manipulate the size of the arc when swinging your hammer and create different knapping results. The tempering and hardness of the hammer head can also make a difference. Too hard, and the head can be too brittle, splinter and slip off the flint; too soft, and there is little impact. A well-tempered hammer head for flint is one that remains intact, and moulds to the flint on impact. Most people will prepare all the flint for a wall or project off site. This reduces risk, and onsite mess.

Depending on the task involved, the quality and source of flint is crucial. This will make a difference in not just colour but also ease of knapping. For any snapped work, chalk-quarried, or gravel-pit flint should be chosen. For gauged flush work or quoins, more hand selection may be required. The challenge for any knapper, whether experienced or novice, is to know which flints are going to do what you ask of them. It is very hard to tell if a flint has an internal thermal fracture. Some more experienced knappers can tap a flint without breaking it open and tell by the sound it makes. There is a ring rather than a thud/dead sound. Some can sometimes gauge by looks and shape. Neither of these assessments are 100 per cent fool proof. The only guaranteed way is to open up the flint, see what you have got and then make decisions about the most appropriate way forward. Again, this will largely depend on what you are trying to achieve. Snapped work is as simple as putting aside what has a good face and an acceptable bed depth. The most efficient way forward for gauged work or quoins is a vetting process whilst knapping. If you open up a flint and it appears to have the potential to be a quoin or an unusual gauge, it can be worth putting aside for working on at a later date.

The hardest part is to know where to start, where to open up the flint and make the first intrusion. Not one flint is the same, so it is not as if you are striking the flint in the same place every time. However, after a little practice a pattern will form and 'reading a flint' will become easier and more predictable. Due to the formed nature of chalk quarry flints, they tend to be more varied in shape and possibly easier to use for a beginner. Gravel pit flint and sea cobbles have become rounded by river or sea movement and therefore give fewer indications of where to start. Listed below are a set of steps that will aid the knapping process. It is not just a question of closing your eyes and wishfully striking the flint, or just hitting the flint harder if it did not break open the first time.

1. Place the flint in your hand and study it. Look at the size and the shape. If size is important, is the flint large enough to match your requirements. Does the flint have a saddle or a shoulder? These are areas of the flint that have slight concave dips. These are the areas that will aid the opening-up process and produce a clean break. Hitting the flint on a convex lump will create resistance and bruise the flint. This will essentially do very little except frustrate you and create a higher risk of producing unpredictable splinters or shards.

2. Wearing the correct PPE, either sit down and place the flint on your thigh or, standing up, hold the flint on the side of your hip. If you are holding the work on the thigh, padding can be useful and help reduce impact. Most knappers will use a leather flint-knapping strap. The layers of leather on these straps hold the flint in place, reducing impact and the risk of being cut.

3. Hold the flint firmly in one hand and tilt the flint away from you. With the other hand hold the hammer and strike the flint at a 45-degree angle. This is called the 'striking platform'. It is not about strength and the power of the blow, it is more about accuracy, where you hit the flint and with what part of the hammer. This is called the striking platform. Try to slightly tilt the hammer so that the impact is on the edge or blade of the hammer rather than the full face. The arc of the hammer movement will change the strike strength rather than the speed of the movement. Try to keep the strike of the blow at a tangent to the tail and body of the flint. This will produce a cleaner cut and the maximum gain from the flint. Listen to the noise of impact. After time, you can hear the difference between a place that will produce a clean cut and one that will produce resistance. Adjust accordingly.

4. Once the flint has 'opened up' and created a new platform face, assess where to go forward. For random snapped work it may be enough to stop there. Assess the flint for thermal fractures and fissures. These are stress marks and weaknesses that may indicate how predictable the flint may be. Ignore them and the risk of unintentional breaks may occur. If all appears well and depending on requirements, it might be possible to work and reduce the flint further.

5. For gauged work, you can either measure and mark with a ruler or place a template on the flint and mark the face. If working on a gauged or shaped flint, once you have marked the face, hold the flint facing up and strike the face of the flint using the same hammer technique as

'Reading' the flint.

Direct percussion using a hammer.

breaking it open. This is called trimming. Often it is completed with a smaller hammer, or one with a flat blade rather than a rounded head. Gradually work towards the drawn line with firm strikes. Tilt the flint and always strike at a tangent to the line. The angle of the flint and the position of the strike will also change the flake size from short oblique flakes to long and thicker intrusive flakes. On completion of the first trim, turn the flint 90 degrees and work towards the line again. Complete all four sides until the flint face has all four edges. Remember, to gain the maximum form, the flint most gauge-worked has the same flint height, but with varying lengths.

6. Copper rods are often used for finer work, particularly when working quoins. On impact copper tends to grip and mould around the flint rather than slide or slip off the point of contact. It is hard to pick up a flint in raw form and know that it is going to be a quoin. Most quoins' blanks are put aside when working snapped flint.

Patterns and Inclusions Found in Worked Flint

Every time a new platform face is exposed, it inevitably will provide something of interest and often something unexpected. Occasionally it may be a fossil or quartz crystals. Often it can be pattern marks, or curious flint shapes created during the knapping process.

Flint works pyramidally. When striking the flint, the fracture heads at a 45-degree angle. When the flint is struck and runs out of inertia, it becomes a ripple. There is less power in the momentum and therefore the shape changes. The last of the ripple becomes a hinge or step fracture.

A flint face on impact with a hammer or sharp object can be fractured conchoidally. This means after impact has taken place smooth shell-like curves can form. On the point of percussion you will also find conical fractures. These cones of percussion will vary depending on the area of impact. Fully formed cones are sometimes known as flint nipples. These features that have been created from a sharp blow point to the flint face are often caused by impact

Fossil intrusion in flint.

during transportation. This might explain the high numbers found on historic buildings, where the flint would have been transported to site by horse and cart and over rough roads. Once exposed to freeze thaw, these areas can often laminate, causing thin layers of flint to flake off.

The best way to learn how to work flint is to practise, observe and be patient, and practise again. Working on a single piece of flint can be like having a conversation or dialogue. You ask questions of each other, and neither knows where the conversation is going to go. It can sometimes feel like flint is working against you. Often frustrating, it can definitely appear like it has a sense of humour, but it will always give you options; you just need to work with it, not against it.

A cone of percussion and lamination.

Chapter Eight

Building a Wall

I am a great believer in both process and content. You can (and should) read this section on how to actually lay flint; however, it is important that you accompany the learning with practical experience and experimentation. It is challenging to put into words a process that is very visual and often intuitive. This section is about the physical act of laying flints. This applies to many situations, whether it be rebuilding a section of a flint wall or building a new flint wall from scratch. There will always be some variations to laying, depending on the style and finish, but the following good practice should be applicable to most styles and finishes.

Setting out and planning has already been briefly discussed; however, it cannot be reiterated how important the preparation process is. Once the correct preparation has taken place, you are ready to start laying.

Project and Site Preparation

Paperwork

Before any work starts it is very important to make sure you are approaching the project correctly. This will vary from project to project. Depending on the size of the project or level of work involved, do you need professional advice (a structural engineer or flint specialist)? Do you need to involve the Local Authority for planning permission, listed building consent or building regulations? Do you need any party wall agreements? These are all questions that are important to ask before any work commences.

Floor protection

Good site management and a clean working area are fundamental to a successful build. Try to be organised and have separate areas for mixing and flint laying. Before laying, decide if the floor needs to be protected. When possible, it is highly recommended to lay boards down. Ideally, boards can be placed on top of a plastic sheet or tarpaulin for extra protection. It is very hard to keep your materials clean if you are working directly off a rough surface. Flint shards can scratch floor finishes, lime can kill grass and foliage and lime mortar can stain. When you are laying flints, make sure you keep the working area in front of the structure free from debris or unwanted materials. Not only can they be a trip hazard, but it can be exhausting walking and working on an uneven surface for long periods of time.

Area management

It is also important to prepare the area where flints are about to be laid. First, you must make sure that the surface you are laying on is free from dust, dirt or loose material. Cleaning this area can be accomplished with either a trowel, a brush, an air compressor or water. Depending on the situation, it is not uncommon to use a combination of all these items. Using a trowel and stiffer brush will remove the larger and more stubborn particles. A softer brush and water will remove the finer particles and any dust. Using water can be beneficial for a number of reasons. Not only can water help to remove dust and finer particles, but it can improve the bond between the new mortar and the old mortar or base substrate. In some instances, it will also reduce the speed of the mortar curing. This is a very critical part of the laying process (*see* the section on mortar). Mortar curing can be particularly problematic when laying on a dry wall, laying against a dry block backing substrate, and laying up to or around bricks.

Material care

In addition, try to keep each element dry, clean and separate. Where possible, sand should be delivered in 1-ton bulk bags or sealed 25kg bags. This is because loose sand can easily get contaminated with soil, dirt, groundwater salts or other unwanted materials. Cover and protect the sand from moisture to reduce bulking and mixing issues. Lime should always be kept dry and protected from the elements. Hydraulic lime should be stored in a well-ventilated area and raised off the ground to reduce risk of contact with wet or damp ground. Quicklime is reactive so should be stored in airtight containers to prevent carbonation. Lime putty should be stored in sealed tubs with a sufficient layer of water to prevent carbonation. Flint can easily be coated in mud or unwanted mortar, especially when your flint is being delivered or stored. You should consider keeping the flints stored in separate piles. If you are building a double-skinned, freestanding wall, it may be sensible to have piles of flints on both sides of the wall. This is because once the wall starts going up, it will get harder to move the flint over the wall. If it is possible, keep the flints you are using dry. Wet flints can be difficult to handle, as they are hard to hold when knapping or laying.

Laying Flint

Freehand

The flint is laid one by one balanced on a horizontal bed of mortar, with a vertical mortar joint between each flint. The size of both the horizontal joint and the vertical joint will depend on the style or effect you are trying to achieve. The course above should be laid on the subsequent completion of each course. Any vertical or horizontal line can either be checked by eye, string line, level or straight edge. Laying freehand is essential if the wall or structure is curved, or if there are any specific details (such as flint size gradation or colour change).

Freehand definitely requires the most skill and experience from the layer. If it is your first attempt at laying flint, it may be advisable to have a few practice runs of laying a front/vertical board or some stacked blocks. This can be a good way to gain confidence and improve your ability as a layer. You might want to try this a few times, reusing the same mortar. If it is possible, try to use different flints when you are practising. This is because it can become too predictable laying the same flints because you know which combinations work!

Different methods of laying flint

There are numerous flint wall processes depending on the style. One of the hardest instructions to provide in word form is to explain when the mortar joint in pointing or laying might be ready for working to achieve any final finishes. The timing can be influenced by many factors. These may include type of lime used, strength of lime, mortar proportions, laying style, thickness of wall, substrate build up, the time of year, the aspect (north or south facing) and the atmospheric conditions (in particular wind and sun).

Flint laying by freehand

1. *Planning*

   The planning of the flint laying is important. This will be subject to a number of differences, the task involved, the number of people laying, ability and speed of laying, the atmospheric conditions – to name just a few. A balance should be struck between laying an area so that height and weight is not an issue, to keeping joints between different days' work to a minimum. For example, it is better to lay less distance and more vertical courses than one long horizontal course. This will make the wall more structurally sound, reduce the opportunity for

The key objective is to have a flat vertical plane. Try making sure that the top of the flint is always on the vertical plane. This should be the highest point of the flint. If there is a higher point within the mortar joint, it may not be possible to lay the next course of flint into the desired position. Therefore, once the above horizontal mortar bed is in place it may obscure the raised flint element and give the impression of a large mortar joint. It may be necessary to remove the offending part of the flint, making it usable. Other common mistakes are to lean the flint in at the top. This will result in a poor platform for the next laid course.

Once you are confident with your flint selection, try holding it up to rest against the palm of your left hand to see if it balances. The shape may be correct, but this process may indicate if the flint will sit well in the mortar bed. It is important that the depth of the flint is no less than 75mm. This will vary from flint style to flint style. However, it is important to remember that if it is any thinner it may not take the weight of the flint courses to be laid above.

7. *Laying the flint*

Once you are happy with the flint selection, firmly push it into place. The initial placement is important, as it will provide maximum bonding. It will be possible to make minor adjustments later in the process. A firm push should squeeze the new mortar around the new and adjacent flint. At this stage do not worry if there is excess mortar on the front. However, if there are gaps in the mortar joint adjust the cutting progress and amount of mortar for the next joint. There may be a temptation to pat and adjust the mortar to change the position or to consolidate. There will also be an opportunity later in the finishing process to add or consolidate the mortar joint. Any overworking can result in slumping and losing adhesion. This is when the mortar joint falls away from the top

edge of the joint or pulls the flint forward. There may also be a risk of closing the mortar joint. A closed joint is when the fat (finer particles) are drawn to the surface, trapping moisture and delaying the setting time.

Once you are happy with the position of the flint, complete the trowel cutting process and select enough mortar for the vertical joint. These are known as the perpend, perp or sometimes cross-joints. With the vertical joints, use a secondary swift sliding motion at a tangent to the mortar bed. Then repeat the flint selection and flint laying process again. Once at the end of the fresh horizontal bed, prepare another 300–500mm bed, and so forth.

At the end of the day, always leave the top of the joint angled. This will create maximum surface area for mortar adhesion between each day's laying, as well as protecting the joint from potential shrinkage between the different stages of mortar application.

When laying flint, a good choice is one that correctly fills the space, which also has sufficient depth to stay in place and support the next course. Whenever possible stagger the vertical joint and avoid flint directly stacked on flint. Regularly step back and look at what has been laid. Check for consistency and bond. Look at the bigger picture, as so often things can be missed if you are preoccupied with a limited working area. Inspect what has been laid after each course has been laid. Now is a good time for any change in flint selection, not after the course above has been laid.

8. *Checking the line and plane*

Any vertical or horizontal line can either be checked by eye, string line, level or straight edge. Laying freehand can be critical if the wall or structure is curved or if there is any detail such as flint size gradation or colour change.

The line should be set up so that the front top of the flint is touching the line. How precisely you follow the line will differ on the laying style. For example, it is crucial to be as precise as possible if you are laying square gauged flush work with 5mm joints. Any variation off the line will disrupt the coursing or create too much variation in joint size. On the other hand, for a coursed field wall with 25–50mm joints between horizontal flint bands, precision is not as important. However, overall, it is preferable to be as accurate as possible during laying and uphold good practice.

There are a few common issues to be aware of when using a string line. If there is not enough tension in the string line or the distance between the two end profiles is too great, it can create line sagging and therefore an incorrect guide. To remedy this, either pull the line tighter or reduce the length between profiles. This should give you a more accurate reading. It is also important to check the string line regularly. This is especially important if you are using an irregularly shaped field flint, which can push against the line and give an incorrect reading. To address this issue, you can set the string line 5mm above the required course level. This height can be even greater when laying random work. This would be when achieving the correct line rather than height is the main objective.

Good communication and patience are necessary when multiple layers are using the same line. This can create continual movement or fluctuation in the position of the string line. Whether laying flint on your own, or with more than one person, the key is to doubl-check the flint is in the required position when you complete the total course. This means any adjustments or changes can be made before the next course has been laid.

In certain circumstances or with particular laying styles, it is not possible to always use a string line or straight edge. When the wall is curved it is possible to use or make a forma to

work to or hold up after each course. Laying a single-skin random-style wall against a solid backing substrate, it is possible to measure the distance between the front face to the substrate and the front top of the flint. As long as the substrate is even and level, it should be possible to obtain a consistent measurement reading and therefore create an even finish. This can be done with a tape measure; however, for speed and ease it can also be achieved by marking the required depth on your hammer shaft with tape or marker pen. Many flint workers set sawline marks on their hammer shafts (normally at 100mm and 110mm). However, it is still important to occasionally check that the saw-mark measurement is still accurate, as a wooden hammer shaft can reduce in length over time and with use.

The use of a straight edge will also help in retaining even coursing and a flush flint plane. This is often used in conjunction with a string line. It can be particularly useful when laying field flint and can be used to either pull flints to the front edge or for levelling off the tops of a laid course. Always keep an eye out that the straight edge has not been damaged (easily done when left onsite or sitting around), as this may result in a poor edge and an incorrect reading.

An alternative to an expensive straight edge can be a straight piece of timber. This can be

Laying a random field flint wall by freehand.

a crude but effective tool. One advantage of using timber is that it can be carefully struck with a hammer when pressed up against the wall to ensure an even and consistent surface. However, this should be a last resort and completed with caution. Although it can push in flints that are protruding too far out, if it is done too violently or soon after the flint has been laid, it can damage the wall.

9. *When to lay the next course*

With good planning one should be able to lay flint all day. As to when is an optimum time to lay the next course, it is down to allowing sufficient curing time of the previous flint course. This will depend on many variations including: atmospheric conditions; flint style; aggregate size and mortar type; depth of flint and bed; thickness of wall – and whether it is freestanding and double skinned or single skinned; and type of substrate.

Before the next course is laid it is important to infill behind the flint. This is sometimes known as backfilling. On a double-skinned wall this can only happen once the flint courses have been laid on both sides of the wall. If only laying a single-skinned wall against a backing substrate, the backfilling can happen as soon as the flint course is ready to take additional weight. Again, there is a balance of waiting for the flint course to have cured sufficiently enough to support the backfill, but not too set to reduce preferred bond of wet on wet.

To complete the backfilling, first lay a bed of mortar in the cavity behind the single-skinned course, or between the two courses of a double-skinned wall. Without dislodging any flint, be sure to push the mortar in sufficiently to engulf the rear faces of the laid flint. Then place waste flint or old mortar lumps (if rebuilding a wall) into the fresh mortar. Work with the shape of the cavity. Try whenever possible to interlock the cavity with suitably shaped overlapping waste material. If selective in choice of shape, this can be particularly useful in bonding two outer flint skins together. Try to vary the sizes of infill material. Be careful that the material used is dust free, and not to over fill. The infill pieces should not go higher than the laid course, as they might infringe on the next course to be laid. Avoid, when possible, for the infill pieces to be touching each other. Once sufficient infill is in place, fill additional mortar on top. Push down mortar to engulf and bond all the infill pieces, not only together but to the rear of the flint course. Again do not overfill, as this may impede the laying of the next course of flint.

Once the cavity has been filled the next course can be laid, and so forth.

10. *Mortar compression*

Whilst the mortar is completing its initial set, you may need to compress the joint (often termed 'driving in the mortar'). You can do this using a natural bristle brush (called 'churn brushes') or your trowel handle. If you have a choice, the long handle brushes provide maximum movement flexibility and greater pivoting when performing the compression action. This action has multiple purposes. It drives the mortar into the joint and reduces any risk of hidden voids and cavities, it increases surface area to encourage curing, and it lets the joint breathe.

If the joint is compressed and pushed in with a trowel, there may be a risk of closing the mortar joint. A closed joint is when the fat (finer particles) are drawn to the surface, trapping moisture and delaying the setting time. This may result in future issues by reducing the transfer of moisture, increasing the chances of water saturation and thus making it more vulnerable to freeze thaw. The timing of the compacting, as with any of the final pointing techniques, is critical. If the joint is compressed too early the mortar can get trapped within the bristles and get pulled out on the withdrawal action of the brush.

11. *Joint protection*

While the mortar is setting, but before any required final finishing techniques are completed, it may be necessary to protect the new pointing. Depending on the weather, protection may be needed from frost, direct sunlight or wind. All these issues can be minimised by the use of hessian. Whenever possible try to use sack hessian, or one with a tighter weave. This type of hessian holds the moisture better and reduces risk of damage to work if direct water spraying. If you are protecting a structure from the sun and wind, it may be necessary to lightly spray the hessian covering with water. Be cautious to not over-soak the hessian as this can result in over-saturation, slumping or washing of the new pointing.

Finishing Techniques

There are numerous pointing finishes that are achievable. When possible, good practice would be to match existing textures and profiles. Unfortunately, repointing is often misinterpreted as an opportunity to draw attention to the mortar or show off trowel skills rather than remain secondary to the flint.

Depending on the flint style, there are numerous flint wall processes. One of the hardest aspects to communicate via writing is when you know the mortar joint in laying might be ready for you to work on any final finishes. The timing can be influenced by many factors. These may include the repointing style required, flint style, mortar used, depth of wall, type of substrate, weather conditions and temperatures, aspect of face (direct sunlight or wind) and amount laid (retained moisture content).

Working the mortar too early can scar the mortar or cause slumping, and leaving it too long may result in the mortar being too hard, therefore making twice as much work, or not being able to control the final finishing as you may have hoped.

Brushed flush pointing

This is probably the most common joint finish. It applies when the mortar joint is recessed on a parallel plane from the front face of the flint. The depth of the recess may vary subject to choice or any matching. A balance must be struck between aesthetics and the retention of a full joint for longevity. Most brushed flush pointing will be sufficiently recessed enough to reveal the edge and shape of the flint. This will reduce the risk of trapped water and draw attention to the flint and not the pointing. As mentioned above, the optimum time to complete this process will be subject to various factors. There are numerous methods to achieve a flush finish. After the flint laying process has been completed, first compress the mortar around the flint. We use either

An example of the mortar joint distracting from the flint.

An example of poor repointing.

a gloved hand for flexibility or the handle end of a wire brush. It is important to seal the joint around the flint to prevent water ingress and enhance stability. After completing this process, scrape excess mortar off with the edge of a trowel. Then, by using a wire brush, remove more of the mortar face; at all times, attempt to use circular motions and avoid deep scarring or digging out of the mortar, as these will cause shadows and an unsatisfactory finish. Be methodical and gradually work around each flint. It is sometimes helpful to start with a wider-gauged wire brush and then select a thinner-gauged brush to complete the finer work around the edges of each flint. If the mortar sticking into the wire brush or process is causing too much scarring of the joint, the mortar probably needs additional time to cure. If the removed mortar is falling out of the wire brush and to the base of the wall it is a good time. Step back and assess the completed brushing. Is it even and consistent? Have any areas been missed? Are you revealing enough of the flint edge? Once the wire brushing is complete, use a natural bristle brush to brush over the whole wall. As with the wire brushing, try to complete in circular motions. This process eliminates any wire-brush marks and accentuates aggregate.

Trowelled flush pointing

This technique is commonly used on field walls and garden boundary walls. It is similar to brushed flush pointing in that the mortar joint is parallel to the face of the flint, except that there is no to little recess. This style is more a case of function over aesthetics, with mortar sometimes covering the edges of the flint. It provides maximum longevity in terms of depth of mortar and weathering. The mortar is compressed with the back of a trowel in a sliding motion. Each action should compress into the previous action, resulting in a consistent overall finish. Step back and assess the completed trowel marks. Is the joint full and evenly consistent? Have any areas been missed? Does it require additional mortar to achieve consistency? Apart

Double line pointing on a coursed field wall.

from the aesthetics of sometimes more mortar than flint, the issue with a trowelled flush finish is that the metal trowel action can draw the fines to the surface, not only hiding any aggregate content but increasing frost damage.

Line pointing

This technique is commonly used on field walls and garden boundary walls and normally used in conjunction with trowelled flush pointing. Complete the same finishing process as specified in the trowelled flush pointing section above. This line creates the optical illusion of a tighter bed joint and provides more texture to an otherwise flat and smooth surface. Then, using a straight edge, lightly score a line in the middle of the horizontal bed. Start from the top of the wall making your way down. This should aid the process of guaranteeing a series of even parallel lines. This should be completed left to right. Be firm and consistent on the pressure of the trowel.

Galleting

This is one of the most labour intensive of all the laying styles. The first stage is to prepare all your gallets ready for use. The colour, shape and length will depend on the galletting style that you are attempting to emulate. Whichever colour or length is to be used, it is important to make the shape of the gallet as thin as possible (at least on one edge). This should

Gallet placement in a floor.

Gallet placement.

be relatively easy to achieve with a well-selected flint source that produces clean and predictable flakes. If the flakes on the reverse side to the strike are too thick, try changing the angle of the flint.

Once the gallets are prepared, the laying can commence. Lay the flint according to the desired style. There are often two differences compared to regular/normal laying. The first difference is the choice of flint-laying aggregate. Often a much finer aggregate is used in the mortar to aid the process of pushing the gallets into the mortar joint. Any large aggregate particle sizes can impede gallet placement and reduce suction. The second difference can be mortar joint size during the initial laying. This will be led by the galleting style required. If the mortar joint is too small, not only will it limit the amount of gallets that can be used, but it can also limit the gallet placement. In particular, the ability to overlap the gallets to achieve visual movement and flow is highly desirable.

With galleting it is important to work a limited area at a time. The gallets are inserted into the mortar joint section by section during the working day. Lay an area more than two or three mortar joints high. If laying random style try to achieve 300–400mm in height. This will allow the opportunity to create movement in the galleting pattern. Galleting only course by course can reduce this opportunity and impede on the overall effect of consistent flowing movement. Depending on atmospheric conditions, the mortar used and the

depth of the wall, the gallets should be placed into the joint after an initial set but whilst the mortar still has some moisture. Once a specified area has been laid, cut around the edge and shape of the flint to achieve a recessed even joint. Then using a natural stiff brush (churn brush) tamp the mortar joint. This action has a dual purpose of making the mortar joint aesthetically consistent, removing any trowel marks, and also opening the mortar joint to aid the process of pushing in the gallets. As mentioned above, the optimum time to complete this process will be subject to various factors. Too early and you will reduce the stability of the flint (resulting in possible slumping), and too late will make it too hard to push the gallet into the mortar joint.

Once the mortar joint has been prepared the galleting can begin. The style of galleting can vary immensely. Density, angle and depth are just a few variations. Most gallets will be pushed vertically or on the diagonal. This will aid the flow of water off the wall and limit any trapping or water ingress. Transport and hold the gallets in either a tray or bucket. Make sure that you have a variety of sizes and shapes. Start at the top of the laid section and insert the gallets. Firmly push in each gallet (thin side into the wall) to the desired depth. Ideally, there should be a minimum of a third of the gallet into the mortar. Avoid any position replacement of the gallets. Once pulled back out there is reduced suction with the mortar. Systematically work your way down the wall. Start with larger gallets first and then infill with smaller gallets. Try and get the gallets to work around the shape of the flint, using curved or diagonal gallets. Also, when possible try to overlap the gallets. Both of these actions will help provide movement in the galleting effect. Unless the effect desired, butt jointing or pushing in the gallets too vertically and horizontally can make the overall effect very static.

As with flint laying, if more than one person is galleting, it is crucial to vary and mix up who is galleting where. Galleting placement is surprisingly individual. I would liken it to handwriting, as the selection and placement of each gallet can really be specific to each person.

Recessed sprayed

This technique has become popular in more recent times to achieve a more organic look when using field flint and hydraulic lime. It is a good way to really accentuate the aggregate and create a weathered look. However, a warning: it is not suitable for all limes, as by the process of saturating the wall it can cause freeze-thaw issues in the winter – and it can cause a mess. After the flint laying process has been completed, first compress the mortar around the flint. We use either a gloved hand for flexibility or the handle end of a wire brush. It is important to seal the joint around the flint to prevent water ingress and enhance stability. Be methodical and gradually work around each flint. Is it even and consistent? Have any areas been missed? Once this process has been completed, by using either a handheld water spray pump or a hose with a diffused spray attachment, spray the mortar to achieve the desired effect in terms of depth, joint shape or aggregate exposure. Essentially, by using a spray technique you are removing the binder. Start at the top and work your way down. The optimum time to complete this process will be subject to various factors – atmospheric conditions, depth of wall, mortar type, and so on. Normally we would spray at the end of the day. If you spray too early there is

a risk of over-saturation and destabilisation of the newly laid flint. After the spraying of the newly laid area is complete, it will be necessary to spray below to reduce the build up of lime deposit.

Snail's creep

There are two ways to achieve a snail's creep effect. The traditional way of adding on the 'creep' in a finer mortar after the laying is to complete (similar to brickwork tuck pointing), or create the creep during the building process using the existing mortar.

In the traditional process, first complete the total wall to be built. Lay the flint using regular mortar. During the build process, cut around the edge and the shape of the flint to achieve an even joint. Then, using a natural stiff brush (churn brush) tamp the mortar joint. This action has the dual purpose of compressing the mortar and making the mortar joint aesthetically consistent, removing any trowel marks. As with other finishing techniques, the optimum time to complete this process will be subject to various factors. Too early and you will reduce the stability of the flint (resulting in possible slumping), and too late will result in the mortar being unworkable.

On completion of the wall you are ready to apply the 'creep'. This is applied using a much finer-aggregated mortar. From my experience the lime content has also often been increased to improve workability. First,

Heavy snail's creep covering the flint.

dampen the wall down to improve adhesion. Then push in and compress the new pointing mortar into the existing flat joint. This can be completed by either a small pointing trowel, tuck trowel or brick jointer. It is crucial to try to obtain maximum adhesion between the two mortars. The pointing does not have to cover the whole of the laying joint, just the centre of the joint. Once a certain area has been pointed, cut the mortar to the desired angle. This can be undertaken either by a brick jointer, small pointing trowel or, as we do, using bespoke tools depending on the size of 'creep'. We ended up adapting and making tools to achieve this effect as we found that brick jointers were not sharp and angled enough for the task, dragging rather than shaping the joint. Using a small trowel it proved hard to achieve a consistent width and angle creep. Depending on the finish, we have ended up using either a shaped and sharpened copper pipe or a split fork. Try to follow the shape of the flint with curving movements rather than angular movements. The snail's creep is meant to enhance and reflect the natural form, not distract and work against it.

Learning how to complete snail's creep is a very good example of how I have learnt a lot of the techniques for flint laying finishing – by trial and error and learning from others. None of the techniques is the only way to achieve the desired result. Some are adapted and improved. There are a lot of similarities in techniques, but most flint layers will have their own preferred tools or ways to achieve finishes. There are few practical books or leaflets. With the presence of the internet, some techniques can now be shared online. Learning from others can be a great way to pass on skills. A few techniques such as snail's creep I have learned orally from others.

An alternative and more recently popular way to achieve snail's creep is by making the pointing shape at the end of the day similar to many other pointing finishes. I am not sure if this is a lack of knowledge or ability, but I can see the attraction and it reduces the risk of pointing separation that appears to have happened to so many snail's creep walls. Lay the flint using regular mortar. When the mortar has set but

is not unmalleable, cut around the edge and shape of the flint to achieve an even joint. Then, using a natural stiff brush (churn brush) tamp the mortar joint. This action has a dual purpose of compressing the mortar and making the mortar joint aesthetically consistent, removing any trowel marks. As with other finishing techniques, the optimum time to complete this process will be subject to various factors. Too early and you will reduce the stability of the flint (resulting in possible slumping), and too late will result in the mortar being unworkable. Then – depending on the desired finish – as with the more traditional technique, use either a shaped and sharpened copper pipe or a split fork. Sharp brick jointers can be more useful than when adding on 'creep'. Tuck trowels or small pointing trowels can be used but still have the issue of uneven and less consistent creep joints.

Bird's beak or weather struck

The processes for both bird's beak and weather struck (and cut) are very similar. The only difference is that bird's beak pointing is always used if both the horizontal and perpendicular joints are to be cut.

Lay the flint using regular mortar. Make sure that all the mortar joints are full and pay more attention to not breaking the horizontal bed line with a long flint. This may impede an even and consistent finish. Then, using a natural stiff brush (churn brush) tamp the mortar joint. This action has a dual purpose of compressing the mortar and making the mortar joint aesthetically consistent, removing any trowel marks. As with other finishing techniques, the optimum time to complete this process will be subject to various factors. Too early and you will reduce the stability of the flint (resulting in possible slumping), and too late will result in the mortar being unworkable.

To achieve a horizontal and perpendicular bird's beak effect

Take a tuck trowel or small pointing trowel, compress and drag an angled trowel in a downward motion between each vertical joint. For all joints, keep the

Bird's beak effect using sea Flint/cobbles.

trowel at the same angle and the line of the vertical drag the same. These do not have to be vertical, but it helps if they are parallel. Most bird's beak pointing is used in conjunction with 11 o'clock or 1 o'clock tilted field flint. Therefore, the vertical cut will follow the angle of the flint. Once this task has been completed, reverse the angle of the trowel and complete the same motion. If successful, you should achieve a series of regular and even parallel bird's beak joints between all the vertical joints.

Then, using a straight edge, lightly score a guideline in the middle of the horizontal bed. Start from the top of the wall making your way down. This should aid the process of guaranteeing a series of even parallel lines. Then taking a tuck trowel or small pointing trowel, drag the back of the trowel along the guideline. This should be completed left to right. Be firm and consistent on the pressure of the trowel. This can be completed free hand or preferably resting along the top edge of a straight edge. The angle of the trowel tilt will define the angle and size of the

bird's beak. On completion of this, raise the straight edge and complete the same task dragging the trowel along on the bottom of the straight edge. Be careful to follow the guidelines to achieve parallel bird's beak. Try to achieve a consistent and equal angle on the top and the bottom of the 'beak'. The horizontal beak should cut through the top and the bottom of the vertical beaks. At the end of the process, carefully cut away loose mortar deposits that have not naturally fallen away. Finally, when the mortar has cured, use a soft brush to remove any rough lumps or score marks.

Bird's beak effect using sea flint/cobbles

I believe that this is probably one of the most technical of all pointing finishes. Due to the uniform nature of the cobbles it is very unforgiving. Any discrepancy in either the laying or the cutting process clearly shows up. From my experience, often either a finer aggregate is used in the laying mortar, or as with snail's creep, the bird's beak pointing is completed after the wall has been built. Historically, for what appears to be solely for visual effect, this was often undertaken using a dark mortar. The other tendency is for the perpendicular joint to always be vertical and never at an angle.

To achieve bird's beak with sea cobbles and using the laying mortar, just follow the bird's beak horizontal and perpendicular method outlined above. To achieve bird's beak with sea cobbles but using different mortar, first complete the total wall to be built. Lay the flint using regular mortar. During the build process, cut around the edge and shape of the cobbles to achieve a recessed joint. As with other finishing techniques, the optimum time to complete this process will be subject to various factors. Too early and you will reduce the stability of the flint (resulting in possible slumping), and too late will result in the mortar being unworkable.

On completion of the wall you are ready to apply the bird's beak. This is applied using a much finer aggregated mortar. From my experience the lime content has also often been increased to improve workability. First, dampen the wall down to improve adhesion. Then push in and compress the new pointing mortar into the recessed joint; this is best completed by using either a small pointing trowel or gauging trowel. It is crucial to try to obtain maximum adhesion between the two mortars. The pointing needs to fill the whole of the recessed joint. Once a certain area has been pointed, follow the bird's beak horizontal and perpendicular method described above.

To achieve a horizontal bird's beak or weather struck effect only
Then, using a straight edge lightly score a guideline in the middle of the horizontal bed. Start from the top of the wall making your way down. This should aid the process of guaranteeing a series of even parallel lines. Then, taking a tuck trowel or small pointing trowel, drag the back of the trowel along the guideline. This should be completed left to right. Be firm and consistent on the pressure of the trowel. This can be completed free hand or preferably resting along the top edge of a straight edge. The angle of the trowel tilt will define the angle and size of the bird's beak or weather struck joint. On completion of this, raise the straight edge and complete the same task dragging the trowel along on the bottom of the straight edge. Be careful to follow the guidelines to achieve a parallel cut. With bird's beak try to achieve a consistent and equal angle on the top and the bottom of the 'beak'. With a weather struck joint there tends to be less of any angle on the undercut action. At the end of the process carefully cut away loose mortar deposits that have not naturally fallen away. Finally, when the mortar has cured use a soft brush to remove any rough lumps or score marks.

Capping

The finishing or capping of a top of a wall is a very important element to guarantee maximum longevity. If it is a freestanding wall it is important to prevent water ingress and any risk of the two skins splitting and reducing their bond. Depending on location, period of build and status there are numerous ways to cap off a wall. Very popular for boundary walls and field boundary walls in rural locations is to have a haunch capping. This is when the mortar continues over the top of the wall in a curved profile. For aesthetic and strength reasons, sometimes flint is added into the haunch; or sometimes additional aggregate is added. After the Industrial Revolution, the mass production of bricks and improvement of the transport system meant that capping bricks became widely available. These would normally be half-round or triangular (known as saddlebacks). The profiles on both these items can vary in detail and in height. The majority of capping bricks are a width of 225mm. This matches the average width of the top of a freestanding wall. Second-hand capping bricks can be hard to obtain, especially a reasonable number of equal size and consistency. It is possible occasionally to get them from reclamation yards; however, this can prove expensive. It is possible to buy new, though these would normally be in a metric rather than imperial size. Other capping options are brick on edge, stone capping or reconstituted stone. A recent popular alternative is the use of pressed, powder-coated zinc. It is rare to find this in boundary walls, but it is popular in recent domestic new builds, particularly as coping on parapet walls.

Winter Works

Winter works with all limes is always possible within reason. However, it may take longer, require more planning, caution and more flexibility. Failures can happen when concentrating on night-time temperatures rather than selection of lime and reducing daytime saturation of materials and substrates. Damage to mortar and structures happens when freeze thaw occurs. Freeze thaw is when water saturation fills voids and cracks and expands. The initial result may just be fine flaking and surface spalling. A continual freeze thaw cycle of thawing and freezing will eventually damage the structural integrity of a wall.

Many traditional practitioners tend to veer away from air limes such as hot lime and lime putty during the winter, due to the need for good atmospheric setting conditions that enable water evaporation. Just reverting to NHLs in the winter months for the sake of convenience is not the answer. Recent evidence has shown that air-trained mortars may have a slower set, but their pore structure is less vulnerable to frost than NHLs (Wiggins, 2017).

Although air limes tend to have a quicker initial set, the slower final setting times can inevitably reduce the ability to build any reasonable height, especially with wide and freestanding walls. Within reason this can be addressed by keeping materials dry, using dry infill, retaining a dry substrate and good in situ project planning (where possible laying a longer horizontal stretch with less immediate height gain).

The effect of freeze thaw on an historical mortar.

Good practices for working in the winter include:

- Schedule planning. Select projects that involve less exposed environments, less risk of water saturation (such as retaining walls), and smaller-scale projects that are easier to protect.
- Protect materials, structure and substrate from saturation. This will not only reduce the risk of freeze thaw, but also the excessive movement of free lime.
- Be prepared to stop work, however inconvenient the timing.
- Do not increase mortar strength by changing lime or using additives at the expense of matching the existing mortar.
- Be prepared to use additional measures including heat or extra covering materials. Hessian will protect a wall from temperatures but not water ingress and saturation. When possible, use a tighter woven hessian or insulation as well as a waterproof covering. Bubble wrap is popular to use, as its pockets can retain some of the daytime temperatures.

Chapter Nine

Support

Inspection

A proactive approach to inspecting a flint wall will always prolong its life. External boundary walls in particular are often ignored or sit in a queue behind house maintenance work and immediate repair tasks. It might be a combination of not being in the garden for a while and the belief that flint structures require specialist repair work. That is not to mention worries about where you would even find a repair person and the price that they may charge you! Before you know it, one or two flints have fallen out or you cannot even see the wall because it is covered in ivy. These are common reasons for why clients' walls have fallen into such disrepair.

Apart from being of general interest, inspection can also help in the evaluation of a structure's general condition and stability. If there are any issues with the wall, this can help identify remedial solutions. Prompt and appropriate action can reduce further damage and increased costs. If problems have occurred with a structure, it is important to identify the original causes to prevent repeated damage. Therefore, it is recommended that you complete an annual inspection of any flint work, or at the very least an annual glance at your wall! There are a number of features that you should be considering when inspecting your wall, in particular the condition of the wall and the presence of foliage. A correct assessment will result in the appropriate action to take.

The wall condition

I personally have a preference for the phrase 'reading a wall' rather than 'inspecting a wall'. The latter may indicate the current condition of the structure, but it might not always help you understand why (if any) problems have occurred. A few questions when reading a wall may include:

Has the height of the wall been changed?

Raising a wall higher can change the load weight of the structure and cause a number of structural issues. Signs of a changed wall height include a change in laying style, a change in mortar and a change in finishing. Often the original wall foundations have been ignored and not adjusted to the new weight. However, the height of a wall is a crucial part of the foundation dimension calculations. Any height raising must only be completed after careful consideration and further exploration of existing foundation dimensions and soil type.

Have the ground levels changed?

Most walls are designed so that at least one element is below ground level. As well as indicating a change in wall height, differences in laying style, mortar and finishing may also be signs of adjusted ground levels. In particular, differences in laying style, size or material being used is often an obvious signal of ground-level changes.

It is quite normal for larger flints or chalk blocks to be used below ground. This section of the wall is of course not usually meant to be seen, but changes in ground levels over time can expose it. Therefore, any underground soft material that was previously protected from the elements can become eroded and create serious structural issues. In addition, the joint between the foundations (assuming it has one) and the start of the flint work can be very vulnerable. This joint is normally horizontal and represents the change from one material to another. Most walls will start well below ground level to protect this joint and improve aesthetics. However, if it is exposed, it is susceptible to erosion and water ingress: weakening the wall.

Has the function of the wall changed?

Some walls may have originally been built to be an internal wall. Therefore, the previous and current function of the wall is worth considering, particularly for older structures and farm buildings. Indications of the wall's function may include: changes in mortar coloration, the use of lime wash, the use of a softer material or evidence of an old mortar roof fillet.

Have the atmospheric conditions changed?

Many pre-mid-19th century walls appear to have a more solid structure. This might be because of the lime mortar used at the time, milder conditions or better ventilation. Mortar failure is common on internal walls above radiators and external walls around either boiler flues or extraction units. This is because for boiler flues and extraction units, the lime mortar breaks down as a result of extreme changes in temperatures and excess moisture in the air.

Have there been previous repairs?

Although they were probably completed with good intentions, poor existing repairs can cause more harm than good. In particular, the use of cementitious mortar on an originally lime-based structure can cause significant problems. Both experience thermal heave at different rates. Consequently, cracking occurs because the cement cannot move and expand. Though lime-based mortar lets water in and out, cement-based mortar can trap water and cause

dampness. Therefore, as cracks appear water gets in and remains trapped, causing internal damage.

In addition, cement pointing on flints makes it very hard to remove and produce clean flints. When taking down a flint wall that is covered in cement, the reclamation rate is significantly smaller than that of a lime wall. If a flint layer is rebuilding a wall that has been built in cement, it is unfortunately more economical to obtain new flints than attempt to salvage the old materials.

Foliage problems

In addition to asking yourself the questions above, it is essential to assess the structure for natural damage from plants.

General foliage

Is there foliage on the wall? If so, how much? What type? In an ideal world, structures should always be kept free of heavy foliage. Small plants such as lichen or moss will always grow on the structure and cause little harm. However, larger foliage will often cause the breakdown of a wall. This can be because the foliage holds water that causes freeze thaw and the roots and stems can expand to create structure breakdown. In addition, large foliage encourages organic matter and can be so weighty that it pulls a structure over.

Can I grow any plants on a flint wall? It is always a risk that they may grow out of control and damage the wall. However, carefully selected or well-maintained plants can enhance the beauty of a flint wall while causing only limited damage. By choosing 'self-clinging plants' such as Virginia creeper or clematis, the life expectancy of the wall may increase. This is because these plants grow on a sucker system or rest on wires rather than eating into the wall. In addition, small plants can grow on the wall in moderation. Pockets of toad flax or alyssum can visually break up the large expanse of a flint wall while causing very little damage. The three plants that appear to cause the most damage to walls and flint structures are ivy, buddleia and elder. So, beware!

Ivy

Ivy has the potential to cause significant damage to a wall. This can be physical damage from heavy and intrusive vines and chemical damage from leaves breaking down the lime mortar. This is because the decay of organic matter produces humic acid, which in turn can dissolve carbonates (that is, lime mortar). The aggressive growth of ivy can split a wall or dislodge flints. There is often an assumption that this damage will only happen to a structure that is already in a state of disrepair. However, even a structure in relatively good condition can be vulnerable to ivy damage. Flint walls are not always necessarily dense and solid masonry structures. The internal structure can contain small voids within the wall, making it more susceptible to root ingress.

It is essential that the vine does not spread or become top heavy. It is common to rebuild walls where the structure could not cope any longer due to the pure weight of the foliage. This seems to occur mostly in late autumn when the ivy has had a good year of growth and there is an increased seasonal chance of heavy rain and strong winds. The original flint structure is never usually designed to cope with changes in stress and weight load. In addition, established ivy can also hide damage or possible issues with the wall as well as encouraging insects. Although insects are not necessarily a bad thing, both ant nests and woodlice infestations can contribute to the breakdown of a wall. Therefore, building a bug hotel or encouraging insects to thrive away from the wall can be beneficial.

Depending on the extent of the ivy, there are often some solutions for managing it. If the ivy is only on the surface of the structure, cutting the main root stem at the bottom of the wall is an effective tactic. This should be done with an old saw, as you will most likely be sawing directly against the masonry, meaning saw blades can be easily damaged. You can then leave the ivy for four to six weeks, checking that there is no new growth. Please take precautions if you are completing this task with a hatchet or axe. It is very easy for the

metal blade to make contact with flint or aggregate particles causing possible injury (for example, tool bounce-back or eye damage via flint shards). Once the ivy has turned brown and appears to have died, it is possible to peel the growth off the wall without causing too much damage.

If it is not possible to physically remove the ivy, systemic pesticide would be advisable. These pesticides can come in the form of liquid or crystals. It is essential that they are used responsibly and you always follow the manufacturer's instructions and guidelines. Normally when any treatment is being performed, the wall should be protected and not exposed to inclement weather. This is because rain will increase the risk of washing chemicals away and this could put other plants and wildlife at risk. Finally, consideration should also be made for the environment when disposing of excess pesticide.

Things become more complicated if ivy is an integral part of the wall. The only way to remove the ivy is to remove the section of wall it has established itself in. However, financial constraints mean this is not always possible for everyone. If that is the case, and the structure is not at risk of collapse or of danger to others, managing the ivy may be your best option. Regular reduction of growth (and therefore weight) also enables a more realistic visual assessment of any changes to the structure. This option can often buy you some more time in protecting the wall from damage until you have the budget to remove the ivy and rebuild the section of wall.

Trees

Trees can be beautiful additions to open spaces. Personally, I try to remind a client of the time it takes for a tree to grow compared to the time it takes to rebuild a section of flint wall! As with the ivy, managing tree growth is a question of management and common sense. It is never a good idea to have a tree of any size in close proximity to a structure. Depending on the tree type and size, the root system and foliage growth can cause serious damage and destabilise

Before completing any repairs, it may also be beneficial to spend a short time researching and documenting the wall. This little effort will go a long way towards keeping the aesthetics and retaining the local distinctiveness of the structure. Damage can occur when the wrong repair techniques and materials have been used. By using correct methods and materials, you are more likely to prolong the life of the structure.

Repointing Remedial Works

Repointing a flint wall can increase the longevity of the structure. You must first understand why there has been mortar deterioration through a detailed assessment. Weathering, poor-quality historic mortar, poor historical repointing, foliage damage, masonry bees, excess moisture, and excess atmospheric salts (locations near the sea) are all causes of mortar deterioration. Once the structure has been properly assessed, you may need to complete other remedial works before any repointing starts.

Repointing is carried out in a number of stages:

Mortar joint preparation
This involves removing any loose or failed mortar from the existing joint (known as 'raking out'). You should decide what level of raking out is to be completed early on in the repointing process. This can be either isolated or throughout the total face of the structure. Unless the mortar has failed completely, you should leave as much sound historic mortar as possible in the structure. If you find or create a well-designed mortar match, aesthetics should not be an issue. However, it is beneficial to remove all cementitious mortar that might be damaging to the historical structure. Do complete this task with caution as flints may fall out in the process. In most situations, non-mechanical hand-held tools are appropriate for this task. With more delicate flint work or structures that include softer materials

(brick or stone), a hammer and chisel are best. This limits any physical damage to the material or destabilising of flint nodules. On more stable walls, you can rake out with a small hand-held hammer action drill or a mortar raker. This should also be undertaken with caution; however, rarely does serious damage occur to the flint. The smaller hammer drill is best for causing minimum vibration and for manoeuvring the point around the joint. Chisel points and mortar raker sizes and shapes will vary depending on the task or joint size.

Too often new pointing will fail due to the depth and shape of the prepared mortar joint. The basic pointing calculation is that the depth of the prepared joint should be twice the width of the mortar joint. However, this is not always possible or realistic. For example, in some coursed flint work the horizontal mortar joint is already very sizable, so you might be going too far in, gouging the joint and destabilising the flint. Instead, concentrate on the shape of the raking out. It should be of even depth and be squared off. You might pay particular attention to the top undercut edge, as this may be obscured by a flint at standing eye level and overlooked. You must make sure edges are not curved, as these can reduce adhesion, encourage freeze thaw or cause the mortar to fall out on the slip plane.

Mortar joint cleaning
Once you have extracted the majority of the failed mortar, you can remove any final loose mortar remains, dust and organic matter. You should complete this task systematically, start at the top and work your way down. First, use a natural bristle brush to gently dislodge unwanted matter. Then, if necessary you can spray the joint with water, using either a pressure washer or a water hose depending on the circumstances.

It is important to use the least water possible to complete the task. Excess water can saturate the structure too much and cause unwanted water ingress, or flood the base of the wall making messy and uncomfortable working conditions. For fragile structures,

Careful removal of mortar without dislodging flints.

Removal of dust and excess mortar using a brush.

too much water can change the weight load and cause structural movement. Pressure washers can be useful; however, these too must be used with caution. Too much pressure caused by directional spray can dislodge flints and damage the structure. The time of year and climatic conditions can also influence the best instruments to use for washing down. Oversaturation in cold periods increases the risk of freeze thaw. However, saturation in the dry summer months can encourage a slow cure, reducing the risk of mortar failure. It is all about finding a balance.

Pointing preparation

At this stage, the mortar joint to be pointed should be free of dust and dirt and damp enough to create good adhesion. If the old mortar is too dry, the back of the new mortar can set too quickly. This can result in shrinkage or the mortar becoming powder and failing. In addition, accelerated shrinkage at the back will create an unseen fine crack that means poor adhesion, increased vulnerability to water ingress and freeze thaw, as well as separation between the new and old mortar. This can be a major issue when laying against dry substrates (such as different forms of block work) in a new build. The moisture content requirement may vary depending on the time of year and the structure you are working on. It might also

be a continual process as you work. For example, in our practice we often dampen down the whole wall at the start and then dampen down the sections as we work. It is surprising how dry an historic structure can be and how much moisture it can absorb.

Repointing

Tools

Once the lime mortar has been designed and prepared, you are ready to commence repointing. To complete this task you can use anything from a flat tuck-pointing trowel to a 6-inch gauging trowel. I personally do not like curved jointing irons, because they make it difficult to hold the desired amount of mortar and often distribute the mortar unevenly into the joint. However, this does depend on the shape and size of the prepared joint and preferred pointing action.

Which tools you select may be influenced by the mortar design, the type of lime used, the aggregate particle size and the flint work style. The ready-mixed mortar is normally temporarily stored in rubber buckets. To hold the mortar whilst in the process of pointing, most practitioners will either use a mortar hawk or work off a larger trowel. Hawks are sometimes more suitable for horizontal stone or brickwork. The straight edge of the hawk will align well with the corresponding joint. Some practitioners will also use a gloved hand for mortar joints that have a large variation in shape. If you choose this method, then the mortar may have to be worked in the bucket rather than on the hawk to make the mortar more pliable.

Laying and removing

Depending on the depth of the joint recess, you may have to fill up the joint by compacted layering when you are applying the mortar. This means pressing additional mortar on top of an existing layer until the joint is full, or just over full. Before applying additional mortar to the joint, you may have to wait for each layer to have some degree of set. However, do not wait until it is completely dry, as this may reduce the opportunity for the applications to bond to each other.

Try not to overwork the mortar whilst pushing in the joint. The mortar needs to be workable but not too sloppy. Overworking can result in slumping and losing adhesion (when the mortar joint falls away from the top edge of the joint). Mortar that is too wet will also result in a similar problem with excess shrinkage.

The best way to remove the mortar from the pointer or trowel is to press against the flint or previous mortar application in a sliding motion. When pulled back at an angle the mortar should remain in the desired location, and the process can start again. It is important to make sure that there is a good seal around the edge of the flint. Although flint is relatively easy to clean compared with stone and brick, why make your job harder? Try to keep any mortar on the face of the flint to a minimum.

Work on an area systematically. You may do all the vertical joints first and then move onto the horizontal ones, but complete an area of both before moving on to another. If you do not do this, the pointing will set at different rates and weaken adhesion. A slow cure is often determined by the shared moisture content in a large area of pointing. If you are unable to complete all the required pointing in one go, leave the top of the joint angled. This will create maximum surface area for mortar adhesion, plus it protects the joint from potential shrinkage between the different stages of mortar application.

Mortar compression

Whilst the mortar is completing its initial set, you may need to compress the joint (often termed 'driving in the mortar'). You can do this using a natural bristle brush (a churn brush) or your trowel handle. If you have a choice, the long-handle brushes provide maximum movement flexibility and greater pivoting when performing the compression action. This action has multiple purposes: it drives the mortar into the joint and reduces any risk of hidden voids and cavities; it increases surface area to encourage curing; and it lets the joint breathe.

If the joint is compressed and pushed in with a trowel, there may be a risk of closing the mortar joint. A closed joint is when the fat (finer particles) are drawn to the surface, trapping moisture and delaying the setting time. This may result in future issues by reducing the transfer of moisture, increasing the chances of water saturation and thus making it more vulnerable to freeze thaw. The timing of the compacting, as with any of the final pointing techniques, is critical. If the joint is compressed too early the mortar can get trapped within the bristles and get pulled out on the withdrawal action of the brush.

Joint protection

Whilst the mortar is setting, but before any required final finishing techniques are completed, it may be necessary to protect the new pointing. Depending on the weather, protection may be needed from frost, direct sunlight or wind. All these issues can be minimised by the use of hessian. If you are protecting a structure from the sun and wind, it may also be necessary to lightly spray the hessian covering with water. However, be cautious to not over-soak the hessian as this can result in over-saturation, slumping or washing of the new pointing.

Finishing

There are numerous pointing finishes that are achievable. Good practice would be to match existing textures and profiles. All too often, due to poor repointing practice and care, a repointed wall can look aesthetically unpleasing. Unfortunately, repointing is often misinterpreted as an opportunity to show off trowel skills rather than remain secondary to the flint. The 'weather struck finish' is a particularly common example of this.

Depending on the flint style, there are numerous finishing processes. One of the hardest aspects to communicate via writing is when you know the mortar joint in pointing or laying might be ready for you to work on any final finishes. The timing can be influenced by many factors, not just the atmospheric conditions of the wind and sun. Other influences may include flint style, the time of year, the aspect (north or south facing), aggregate size and mortar type, wall build up, including depth of flint and bed, thickness of wall, if freestanding and double skinned or single skinned and type of substrate. Working the mortar too early can scar the mortar or cause slumping, and leaving it too long may result in the mortar being too hard, therefore making twice as much work, or not being able to control the joint finishing as you may have hoped.

With flint laying, nearly always you have to work with the weather and plan your schedule. For example, when possible we often schedule in larger freestanding walls or exposed locations in the summer. During long summer days, whenever possible we will work in the shade at certain times of the day. At any time of year we will often plan repointing on cloudy days. All of these considerations encourage good working practice.

An example of poor repointing.

Two examples of good finishing.

In writing this book I have sought to bring together my research, archive of images and the experiential learning gathered over many decades of working with flint. Although I was soon given the nickname 'the flintman', I didn't realise when first starting out on this path, that I would become so passionate about this often-overlooked stone. However, the more I have explored and experimented, the more it has become clear that what the material has offered to the fields of architecture, construction and design, is a fraction of what it has yet to give. It is a cliché that rings true here – the more I have learned, the more, I realise, I have yet to know.

My intention here has been to convey some of my passion and to spark your curiosity. My hope is to provide you with some information on flint history, usage, techniques, and best practice as a place to start your own exploration. I hope the images will inspire you. My wish is to offer a resource for the continuing use and appreciation of flint. It is part of history, that is presently enjoying a renaissance; and I believe that it has much to contribute to the future.

Flint seedlings in chalk.

Blatchly, John and Northeast, Peter, *Decoding Flint Flushwork on Suffolk and Norfolk Churches* (Suffolk Institute of Archaeology and History, 2005)

Collet, Helene, *The Neolithic Flint Mines of Spiennes* (Institut du Patrimoine Wallon, 2013)

Collins, E.J.T., *Crafts in the English Countryside: Towards a Future* (Countryside Agency Publications, 2004)

Copsey, Nigel, *Hot Mixed Lime and Traditional Mortars: A practical guide to their use in conservation and repair* (The Crowood Press, 2019)

Dawson, Brian, *Flint Buildings in West Sussex* (West Sussex County Council, 1998)

Erlande-Brandenburg, Alain, *The Cathedral Builders of The Middle Ages* (Thames and Hudson, 2005)

Forrest, A.J., *Masters of Flint* (Terence Dalton Limited, 1983)

Hart, Stephen, *Flint Architecture of East Anglia* (Gile de la Mare Publications, 2000)

Hart, Stephen, *Flint Flushwork, a mediaeval masonry art* (The Boydell Press, 2008)

Hislop, Malcolm, *Medieval Masons* (Shire Publications, 2010)

Hogberg, Anders and Olausson, Deborah, *Scandinavian Flint, an archaeological perspective* (Aarhus University Press, 2007)

Jackson, Hazelle, *Shell Houses and Grottoes* (Shire Publications, 2001)

Lord, W. John., *The Nature and Subsequent Uses of Stone Volume 1* (John Lord, 1993)

Lynch, Sean and Satorre, Jorge, *The Rise and Fall of Flint Jack* (The Henry Moore Institute, 2019)

Mason H.J., *Flint, The Versatile Stone* (Providence Press, 1978)

McAfee, Patrick, *Lime Works, using lime in traditional and new buildings.* (The Building Limes Forum of Ireland and Associated Editions, 2009)

Mercer R.J., *Grimes Graves, Norfolk, Excavations Volume I and II* (Department of the Environment, 1972)

Mortimore, Rory N., *The Chalk Downs of Sussex and Hampshire and the North Downs of Kent* (Geologists Association Guide No.74)

National Conservation Centre, *Short Guides to Masonry* (Historic Scotland, 2014)

Orna, Bernard and Elizabeth, *Flint in Norfolk Building* (Running Angel, 1984)

Schofield, Jane, *Lime in Building, a practical guide* (Black Dog Press, 1994)

Shepherd, Walter, *Flint, its origin, properties and uses* (Faber and Faber, 1972)

Skertchly, Sydney B.J., *The Manufacture of Gun Flints* (Originally Published for Her Majesty's Stationery Office, 1879. Reprinted by facsimile publisher.)

West, G.H., *Gothic Architecture in England and France* (G. Bell and Sons Limited, 1911)

Wiggins, D., 'Traditional Lime Mortars and Masonry Preservation', *Journal of the Building Limes Forum* (2017)

aggregate 102–105
agriculture
    and ceramics 47
    environment 81
Aldryche, Thomas 35
America, 'First Great
    Awakening' in 36
Ancient House Museum 21
architecture, flint
    agriculture and ceramics 47
    Church of Holy
        Transfiguration 28–29
    fishing industry 46–47
    flint knapping 44
    geography and
        hierarchy 30–31
    gunflint industry 45
    historical use in 31–42
    ironworks 46
    raw material 30
    structure 27–31
area management 132
Arts and Crafts movement 38
Ashley, Pony 18, 20
Astbury, John 22, 23
Avery, Fred 20

backing substrate 89
'bag lime' 111
'baked bean' flint 60
ball pein hammer 124
banded flint 56–58
Basham, Robert W. 'Bill' 21
Baudre, Honoré 26
Benson, Thomas 23
bird's beak effect 143–145
black Brandon gunflint 16
black flints 51, 114
blown wall 151
blue patination 27
boute coupe 10
Brandon knappers 15–17, 19
brick banding 57
Bricklayer's hammer 124
Britain
    Roman structure in 31–32
    tea and coffee in 22
Bronze Age (4600–4200bc) 12
Bruges 34
brushed flush pointing
    technique 139–140
brushes 126
'bubberhutching on the sosh' 17
builders limes 107, 111
'Building Regulations' 84
'bungaroosh' work 66
Burrell Foley Fischer LLP 43
Bury St Edmunds Cathedral,
    flush work on 36, 37

capping 145–146
carbon-neutral product 105, 106

carstone chip galleting 56
Carter, William 'Billy' 19
'cast' 20
Castle Acre Priory 33
Catholic church of Saint
    Gregory 39
cement 112–113
cementitious-base blocks 86
ceramic industry 21–25, 47
chalk cliff 7, 8
chalk-quarried flint 7, 27,
    94, 100
chequerboard 58
chequered work 58–59
chert 7
chessboard work 57, 58
Church of Holy Transfiguration,
    Great Walsingham 28–29
churn brushes 126, 138, 143
'circular economy' 70
Cissbury 11
Cley-on-Sea, Norfolk 76–77
clinker 46
coal-fired flint kilns 23
coarse sharp sand 105
Coastal Protection Act 1949 96
cobbles 7, 23–24, 46, 53–54, 70,
    98, 144–145
composite banding 57
construction planning see also
    flint wall construction
    freehand method 80–81
    gabion baskets method 87
    method considerations
        87–90
    pre-cast blocks and panels
        method 83–87
    seasonal demands 103
    selecting and sourcing
        materials 90–119
    shuttering method 81–83
    wall types 80
copper rods 124
cortex colours 8
cracks, in wall 150–151
'Creamware' 23
crème de la crème 51
crowbar 17

dark flint 65
Dashwood, Francis 35, 36
dead lime 111
decorative flush work 65–66
decorative work 61–62
diagenesis 7
diaper work 62–63
diggers, safety reasons 17
direct percussion 128, 130
DIY repair 151
double-skinned walls 80
Dover Pharos 32
dry flint 48

duck egg cobble work 53
Dwight, John 22

East Anglian flint 94
Ellman, John 47
'endangered' 44
England
    'Anglican Revival' in 36
    chalk cliff 7, 8
    chequered work in 59
    flint knapping 15
    floors and roads in 69
    jags and spoil-heaps 21
    mines in 11
    Trent and Mersey Canal
        digging 23
English Reformation 34

'faced' wall 80
fat lime 109
field flint 7, 27, 43, 48–49, 55,
    91–92
    and lime slurry wall 82
finer knapping techniques 126
fishing industry 46–47
'fish scale' flint style 46, 47
'fish tail' flint dagger 13
flakers 19
flaking process 18–20
flint
    Anglo-Saxon period 32
    appearance 8–9
    blocks 86
    Bronze and Iron Age
        periods 13
    ceramic industry 21–25
    construction see
        construction planning
    design specifications 90
    formation 7
    geography 9
    glass 25–26
    gunflint industry 13–21
    and horse-teeth floor 78
    in many tongues 9
    mining 17–18
    pre-cast panel 39
    prehistory 10–13
    'silex piano' 26
    silica in 7, 25
    strike-a-light and steel
        mills 26
    types 7–8
    unwanted 11
Flint House, Waddesdon
    Estate 40–41, 55, 68
'Flint Jack', tale of 14
flint knapper/knapping 10, 12,
    15, 19–20, 128–131
    architectural 44
    Brandon as 20
    hammers 125

flint laying
    anecdote 127
    boards and buckets
        125–126
    brushes 126
    freehand 133–139
    hammers 122–125
    mixing mortar 120–121
    planning 133–134
    principal tools for 120–127
    profiles and block lines 122
    straight edges 122
    trowels 122
    water spray and
        hessian 126–127
flintlock industry 15, 23
flint mill 25
    and grinding 23–25
    in Stoke-on-Trent 24
flint nipples 130
flint picking 92–94
flint size gradation 80
flint styles 48–79
    banded flint 56–58
    chequered work 58–59
    cobble work 53–54
    decorative flush work 65–66
    decorative work 61–62
    diaper work 62–63
    field flint 48–49
    floors 69–70
    galleting 54–56
    gauged flush work 51–52
    knapped/snapped
        work 49–51
    quirky 70–79
    quoins 67–69
    relief work 59–61
    rubblework 66–67
    Snail's creep 64–65
flint wall construction see also
    construction planning
    area management 132
    capping 145–146
    cracks in 150–151
    finishing techniques
        139–145
    floor protection 132
    inspection 146–152
    laying flint 133–139
    material care 133
    paperwork 132
    project and site
        preparation 132–133
    repointing remedial
        works 152–155
    support 147–156
    types 80
    winter works 146
flintwork, reading 27–30
floor protection 132
floor stone 18

Folkington Manor in East
      Sussex 52
'folly' flint 60
fossil intrusion 131
found flint 48
freehand method 80–81,
      133–139
free lime 107
'freeze thaw' 30
French technique 19
furlong mills 24

gabion baskets method 87
galleting technique 54–56, 99,
      140–142
garneting/garreting 54
gauged flush work 29, 37,
      51–52, 74
gauging buckets 126
Ghent 34
glass-making industry 25–26
gravel pit-quarried flint/
      gravel-pit flint 7, 9, 27, 29,
      50, 55, 94–96, 100
Greek and Italian Renaissance
      designs 35
Grief, Elizabeth 16
Grimes Graves 11, 12, 16,
      17, 30
Guildhall, Norwich 34
gunflint industry
      Brandon knappers 15–17
      demise of 20–21
      mining 17–18
      during Napoleonic
            period 15
      origins 14–15
      quartering, flaking and
            knapping 18–20
      in Suffolk and Norfolk 45
      working 13–14

hammers 122–125, 128
Harrow Hill 11
Hayward, Philip 16
Henry VIII 35
Heritage Crafts Associations
      Red List of Endangered
      Crafts 44
hessian 126–127
Hindsgavl Dagger 13
Holbein Gate 35
Holy Transfiguration, Great
      Walsingham 28–29
Holy Trinity Church, flush work
      on 34, 37, 62, 63
horse-teeth floor 77–78
hostile environments 108
hot mixed lime mortar 110
House 19, 40, 42
hydrated lime 111
hydraulic limes 107, 108, 133

independent visual
      analysis 116–119
indirect percussion 128
Industrial Revolution 30, 36

ironworks 46
ivy 149

'jag'/cartload 17–18, 21
joint protection 139, 154

Kent, William 60
King's Gate see Holbein Gate
knapped/snapped work 49–51
'knapper's rot' 23
knapping process 18–20

laminar blade core technology 10
'lamp blacking' gunflint 16
lean/thin limes 109
Le Tréport, Normandy,
      France 74
Levallois techniques 10
lime 105–111, 114, 133
lime putty 109–110, 133
lime putty mortar 109–110
line pointing technique 140
lump lime 109, 114

manufacturing techniques,
      changes in 21
marine organisms 7
marine-washed sand 104, 105
matchlock method 15
micro-organisms 7
milk flint 94
mining 17–18
Mohs scale measures 122
Moretti, C. 25
mortar 27, 49
      analysis and design 115–116
      compression 138
      design 115
      joint protection 139, 154
      lime 101
      mixing methods 120–121
      pigments and 113–114
      selection of 100–101
Muntham Court 36

natural bristle brushes 126
natural hydraulic limes
      (NHLs) 106–109, 146
Neanderthals 10
Neolithic flint mines 10, 12, 16
non-hydraulic limes 109

Opus
      incertum 31, 80
      mixtum 31
      quadratum 31
      spicatum 31
Overstrand 70–72

paddle whisks 121
pan mixers 120
paperwork 132
'the pecking hen' 15
personal protective equipment
      (PPE) 128, 129
Pevensey castle 32
phoenix sand 46, 116

pick hammer 124
pigments 113–114
plasterer's whisk 121
plate flints 64
'platform' technique 19
potstones 8, 9
'The Potteries' 22
'potter's rot' 23
pottery industry 23
pozzolans 110, 111–113
pre-cast blocks and panels
      method 83–87
prefabricated flint panels 83, 84
profile 122
pyrite 26

quarried flint 94
quartering hammer 124
quartering process 18–20
quick lime 109, 133
quirky flint styles 70–79
quoins 67–69, 96–99

raking out 152
Ravenscroft, George 25
'reading' flint 135
recessed sprayed technique 142
reclaiming flint 99–100
relief work 59–61
repeated notches 85–87
replica flint tools 11
Roman
      construction method 69
      flint buildings 31–32
Rotherfield Park Estate 82
rubber buckets 125
rubblework 66–67

saddlebacks 145
'sail' 150
salvaging flint 99–100
sand 102
      and aggregates 102, 103
'Scottish decorative pebbles' 96
sea and tidal movements 27
sea cobble flint 96
sea flint see cobbles
sea pears 8, 9
segregation process 113
self healing 106
shafts 17
'shepherd's knee cap' 93
shuttering method 81–83
'silex piano' 26
silicon dioxide (SiO$_2$) 7
Simpson, Edward 14
single-skin cladding 80
site mixers 120–121
skilled knapper 18–19, 96
slaked lime 109
slaking 109
slumping 88–90
snail's creep 64–65, 116,
      142–143
snapped chalk-quarry flint 50
snapped work 49–51
soft bristle brushes 126

sourcing flint 90
Spedding, Carlisle 26
split flint 96
stacked sieve system 117
steel brushes 126
steel mills 26
St Ethelbert's Gate, Norwich 33
St Mary of the Assumption
      church 65
St Mary's, Walsham-
      le-Willows 34
St Marys, Woolpit 34
St Michael Coslany, Norwich,
      flush work on 36
Stoke-on-Trent, flint mill in 24
stone and snapped quarry-flint
      banding 57
St Peter's and St Paul's church,
      Cromer 36, 37, 52
St Peter's church, West
      Blatchington 29–30
straight edges 122
'strike a lights' 20, 26
striking platform 129

thunderbolts 93
thunderstone 93
Toninato, T. 25
trees 149–150
Trent and Mersey Canal
      digging 23
trimming 130
Triumphal Arch, Holkham Hall
      estate 60
trowelled flush pointing
      technique 122, 140

United Kingdom
      ceramic industry 24
      flint 9
      stone tools in 10
unwanted flint 11
unworked chalk-quarried
      flint 89
unworked flint 48

Venetian glass-making 25
vernacular flint 31
Vertue, George 35
void ratio 118

'wall bleeding' 111
water spray 126–127
'weather struck finish'
      143–145, 155
'wedge' technique 19
Wedgwood, Josiah 23
West Dean House, West
      Sussex 68, 69, 71, 77, 79
West Wycombe Park 36, 59
wet environment 108
wet pan grinding method 23
wetstone grinding process 24
Whalebone House 72–73
whole flint see dry flint
wire brushes 126
wool industry 32–34

First published in 2024 by
The Crowood Press Ltd
Ramsbury, Marlborough
Wiltshire SN8 2HR

enquiries@crowood.com

www.crowood.com

This impression 2024

British Library Cataloguing-in-Publication Data
A catalogue record for this book is available from the British Library.

ISBN 978 0 7198 4322 8

Dedication
I would like to thank my family for their encouragement, support and putting up with my endless diversions to document obscure flint buildings, and especially Hannah-Rose for making me sit down and get on with it.

Acknowledgements
I would like to thank the following people for their time, knowledge and suggestions: Hugo Anderson-Whymark, Jonathan Day, John Lord, Bernard Lovatt, Rory Mortimore, and Norfolk Museum Service.

Typeset by Simon and Sons
Cover design by Design Deluxe
Printed and bound in India by Thomson Press